A Guided Tour of Light Beams

From lasers to optical knots

A Guided Tour of Light Beams

From lasers to optical knots

David S Simon
Stonehill College and Boston University, USA

Morgan & Claypool Publishers

Copyright © 2016 Morgan & Claypool Publishers

All rights reserved. No part of this publication may be reproduced, stored in a retrieval system or transmitted in any form or by any means, electronic, mechanical, photocopying, recording or otherwise, without the prior permission of the publisher, or as expressly permitted by law or under terms agreed with the appropriate rights organization. Multiple copying is permitted in accordance with the terms of licences issued by the Copyright Licensing Agency, the Copyright Clearance Centre and other reproduction rights organisations.

Rights & Permissions
To obtain permission to re-use copyrighted material from Morgan & Claypool Publishers, please contact info@morganclaypool.com.

ISBN   978-1-6817-4437-7 (ebook)
ISBN   978-1-6817-4436-0 (print)
ISBN   978-1-6817-4439-1 (mobi)

DOI   10.1088/978-1-6817-4437-7

Version: 20161101

IOP Concise Physics
ISSN 2053-2571 (online)
ISSN 2054-7307 (print)

A Morgan & Claypool publication as part of IOP Concise Physics
Published by Morgan & Claypool Publishers, 40 Oak Drive, San Rafael, CA, 94903 USA

IOP Publishing, Temple Circus, Temple Way, Bristol BS1 6HG, UK

Dedicated to Alee, Marcia, Tina, and Jo

Contents

Preface	ix
Acknowledgements	x
Author biography	xi

1 Introduction: From death rays to smartphones — 1-1
    Bibliography — 1-3

2 Optical propagation — 2-1
2.1 Electromagnetic fields — 2-1
2.2 Helmholtz equation and wave optics — 2-2
    Bibliography — 2-5

3 Gaussian beams and lasers — 3-1
3.1 Lasers — 3-1
3.2 Gaussian beams — 3-5
3.3 Coherent and squeezed states — 3-8
3.4 Optical tweezers — 3-10
    Bibliography — 3-13

4 Orbital angular momentum and Laguerre–Gauss beams — 4-1
4.1 Polarization and angular momentum in optics — 4-1
4.2 Generation and detection of Laguerre–Gauss beams — 4-6
4.3 Optical spanners and micropumps — 4-8
4.4 Harnessing OAM for measurement — 4-9
    Bibliography — 4-10

5 Bessel beams, self-healing, and diffraction-free propagation — 5-1
5.1 Bessel beams — 5-1
5.2 Optical petal structures — 5-5
5.3 More non-diffracting beams: Mathieu beams — 5-6
5.4 Optical tractor beams and conveyor belts — 5-9
5.5 Trojan states — 5-10
5.6 Localized waves — 5-11
    Bibliography — 5-13

6 Airy beams and self-acceleration — 6-1

- 6.1 Airy beams — 6-1
- 6.2 Self-accelerating beams and optical boomerangs — 6-2
- 6.3 Applications — 6-4
- Bibliography — 6-5

7 Further variations — 7-1

- 7.1 Separable solutions — 7-1
- 7.2 Hermite–Gauss beams — 7-2
- 7.3 Ince–Gauss beams — 7-4
- 7.4 Parabolic beams — 7-5
- 7.5 Elegant beams — 7-8
- 7.6 Lorentz beams — 7-8
- Bibliography — 7-9

8 Entangled beams — 8-1

- 8.1 Separability and entanglement — 8-1
- 8.2 Creating entanglement — 8-4
- 8.3 Applications of entangled beams — 8-7
- Bibliography — 8-11

9 Optical knots and links — 9-1

- 9.1 From knotted vortex atoms to knotted light — 9-1
- 9.2 Knotted vortex lines — 9-3
- Bibliography — 9-5

10 Conclusion — 10-1

- Bibliography — 10-2

Appendix Mathematical reference — A-1

Preface

It has been over fifty years since the invention of the laser. Since then lasers have worked their way into every aspect of daily life, from laser pointers and DVD players to eye surgery and tattoo removal. They have also become an essential tool in virtually every area of basic scientific research. In addition to their high intensity and strong coherence properties, the other main defining property of the light produced by a laser is its high degree of directionality: the light emerges as a well-defined beam with a definite direction.

In recent years a number of other optical beams have become objects of study among researchers. Some of these highly-directional solutions to the Maxwell equations were discussed decades ago and then forgotten; others have been discovered for the first time in recent years. These newly fashionable beams have a range of unusual and at times surprising properties. The array of novel features they present have not only made them interesting subjects for scientific study, but are also rapidly leading to a number of novel technological applications.

Many of the beams discussed in this book have never appeared in a textbook before, and information about them is scattered in the original research literature. The goal here is to collect together in one place the basic properties of many of these beams, and to cover them in a more or less unified manner, at a level that is comprehensible to an advanced undergraduate with a background in the physical and mathematical sciences. This survey may also be useful to graduate students and researchers wanting a rapid introduction to the area.

The collection of optical beams discussed here will likely play prominent roles at the forefront of research in a number of fields for years to come, and the hope is that this book will help the reader gain entry into this world of new discoveries.

David Simon
Easton, MA

Acknowledgements

I would like to thank my friends and colleagues at Stonehill College and Boston University for all their help and support over the years, including Professors Guiru Gu, Michael Horne, Gregg Jaeger, Alessandro Massarotti, and Alexander Sergienko. This book grew out of notes for a summer research project carried out by my students Gregory Costello, Mark Hamalian, Praveen Jain, and Michael Maggio at Stonehill College; it would not have been written without them and I would like to thank them for their work, which included writing the program used to generate the images on the front cover of this book. Thanks also to the very helpful people at Morgan & Claypool Publishing and the Institute of Physics Concise Physics program, especially Jeanine Burke, Joel Claypool, Karen Donnison, and Brent Beckley.

Author biography

David S Simon

David Simon received a bachelor's degree in mathematics and physics from Ohio State University, followed by doctoral degrees in theoretical physics (Johns Hopkins) and engineering (Boston University). Originally trained in mathematical physics and quantum field theory, he now works primarily in quantum optics and related areas. He has been the author or coauthor of dozens of papers on topics ranging from the use of supersymmetry in quantum mechanics to the application of quantum entanglement to optical measurement and cryptography. After spending many years teaching physics and mathematics at Nova Southeastern University in Fort Lauderdale, he is currently a faculty member in the Department of Physics and Astronomy at Stonehill College (Easton, MA) and a visiting researcher at Boston University.

IOP Concise Physics

A Guided Tour of Light Beams
From lasers to optical knots
David S Simon

Chapter 1

Introduction: From death rays to smartphones

The foundations of modern electromagnetic theory are based primarily on the work of Faraday, Maxwell, and their colleagues in the 19th century, who pieced together a coherent conceptual framework that could consistently explain all of the experimental results accumulated over the previous century. It soon became clear that optics was a sub-branch of this more general electromagnetic theory, with light explained as a sort of ripple in the electromagnetic field. The coming of quantum theory added additional wrinkles to the story, in particular the idea that the electromagnetic field was built out of indivisible units called photons; but for most macroscopic phenomena, the overall picture was largely unchanged.

All known natural sources of electromagnetic radiation, and most simple technological sources as well, are undirected: the radiation (whether it be microwave, radio, optical, or in any other part of the spectrum) is sent in many different directions simultaneously, weakening as it spreads farther from the source. Each atom in the source acts as a point emitter, sending spherical waves uniformly in all directions, in accord with the classical Huygens principle. If the source is of finite extent, like a star or a light bulb filament, then the radiation patterns of all of the point sources must be added, which leads to the possibility of interference. In some cases the interference may lead to an intensity pattern that differs greatly from the spherical patterns of the individual points; for example, a linear radio antenna will give a characteristic dipole pattern, in which the radiation is much stronger in some directions than others, with greatest concentration near the plane perpendicular to the antenna. However, the pattern is still largely diffuse and undirected in the neighborhood of that plane.

In optics, it has been known for centuries that light can to some extent be coerced into a particular spatial direction or toward a specific spatial location: lenses and mirrors can collimate light into a fixed direction or focus it to a small region in space. Such manipulations are the basis for an enormous number of useful applications, starting historically with the invention of magnifying glasses, microscopes, and telescopes. However, this ability to direct light with lenses and mirrors is strictly limited: both Fourier analysis and the Heisenberg principle guarantee that light beams of finite size that initially seem well-collimated by a lens will gradually spread out, eventually covering a large range of directions again. Similarly, if light is strongly focused by a lens, it will begin to defocus just as strongly once the focal point is passed.

The idea that light could be contained in a highly directional beam without significant spreading goes back to ancient times. The emission theory of vision, which was subscribed to in various forms by both Euclid and Ptolemy, postulated that Aphrodite lit a fire behind the eye when each person was born, and that this fire produced rays of light that were emitted by the eye. Vision occurred when those rays were reflected back into the eye by an object. This theory of vision was not laid to rest until experiments and mathematical analysis done by Alhazen (Ibn al-Haytham) around the beginning of the 11th century, work that was later greatly extended by Newton. The emission theory was based on the idea that each small portion of light followed a path described by a geometric ray. The modeling of light waves by geometric rays remains a useful tool in optics today, and is somewhat vindicated by the fact that light may be viewed as photons traveling in straight lines.

Although all collimated beams will eventually spread, it is in principle possible to make the rate of spreading imperceptibly slow. The possibility that electromagnetic energy could be deliberately collimated into a stable, intense, and highly directed beam that could persist for a long distance without divergence is one that began to be considered seriously in the late 19th century, although it only became a reality with the development of the maser and laser between the early 1950s and the early 1960s. The ray-like behavior of individual high energy particles from the decay of radioactive nuclei had become apparent by the late 1800s, and possibly this was the initial stimulus for mentions of directed-energy weapons and death rays that have been found as far back as the 1870s. Such weapons were discussed seriously by a number of scientists and inventors by the early 1890s [1]. With the introduction of Martian heat-rays in *The War of the Worlds* (H G Wells, 1898), the idea of ray guns became a staple of science fiction and has remained so ever since.

Fortunately, in real life highly directed energy beams have so far been used primarily for more constructive and peaceful purposes. Variations on the basic theme of the most common such beam, the laser, will be the main focus of this book. It is hard to overestimate the influence the invention of the laser has had on life in the 21st century. It plays an essential role in numerous medical applications (microscopy, eye surgery, and tumor removal to name just a few), that have improved or saved millions of lives in recent decades. Current areas of research promise to greatly expand the role of lasers in the medicine of the future. One such area is the field of *optogenetics*, in which the functioning of individual neurons or other cells in living tissue are controlled by application of intense light pulses [2]. Lasers have been instrumental in many other areas of scientific research, from the control of chemical processes to the recent discovery of gravitational waves created by colliding black holes. They are ubiquitous in more mundane areas of life as well, from DVD players and barcode readers to smartphones and fiber-optic communication networks.

A comprehensive overview of the laser's history and applications can be found in a recent volume celebrating the fiftieth anniversary of its invention [3].

Aside from a brief section in chapter 3, we will not be concerned here with the workings of lasers as devices; many excellent references already exist for that, including [4–7]. Instead, we focus here on the beam itself. The standard laser beam is discussed in the next chapter, and then the remainder of the book surveys a number of more specialized variants on the laser beam. Most of these have only begun to be studied in the past few decades, and some were not even suspected to exist until a few years ago. These new beams often have a range of highly counter-intuitive properties: they can twist like corkscrews, propagate without diffraction, bend into complicated shapes, or heal themselves after being disrupted by an obstacle in the beam path. By interfering several beams, it is even possible to create knotted and linked lines of darkness embedded in a surrounding field of light. The applications of these beams are as diverse as the beams themselves: in addition to many uses in microscopy, they have served as optical tweezers and wrenches to manipulate nano-scale particles, or as miniature-sized versions of the tractor beams famous from Star Trek. They also play a prominent part in recent experiments on quantum cryptography and quantum communication. In the near future, specialized optical beams are poised to potentially play a role in the development of quantum computing.

The goal of this book is to provide a quick introduction to these topics at a level that should be accessible to advanced undergraduates in physics, chemistry, or engineering. The primary prerequisites are a basic familiarity with electromagnetism and optics, and a rudimentary knowledge of quantum mechanics.

Bibliography

[1] Fanning W J Jr 2015 *Death Rays and the Popular Media 1879–1939: A Study of Directed Energy Weapons in Fact, Fiction and Film* (Jefferson, NC: McFarland and Co)
[2] Fenno L, Yizhar O and Deisseroth K 2011 The development and application of optogenetics *Ann. Rev. Neurosci.* **34** 389
[3] Bretenaker F and Treps N 2015 *Laser: 50 Years of Discovery* (Singapore: World Scientific)
[4] Siegman A E 1986 *Lasers* (Mill Valley, CA: University Science Books)
[5] Svelto O 1989 *Principles of Lasers* 3rd edn (New York: Plenum)
[6] Saleh B E A and Teich M C 2007 *Fundamentals of Photonics* 2nd edn (Hoboken, NJ: Wiley)
[7] Milonni P W and Eberly J H 2010 *Laser Physics* (Hoboken, NJ: Wiley)

IOP Concise Physics

A Guided Tour of Light Beams
From lasers to optical knots
David S Simon

Chapter 2

Optical propagation

2.1 Electromagnetic fields

Light is ultimately a propagating wave of electromagnetic energy, so it is appropriate to begin by reviewing some relevant information concerning basic electromagnetism and optics.

Classical electromagnetic theory is based on the Maxwell equations, compiled by James Clerk Maxwell in the early 1860s. In modern notation and MKS units, the differential form of these laws is given by

$$\nabla \cdot \boldsymbol{E} = \frac{\rho}{\epsilon_0} \qquad \nabla \cdot \boldsymbol{B} = 0 \tag{2.1}$$

$$\nabla \times \boldsymbol{E} = -\frac{\partial \boldsymbol{B}}{\partial t} \qquad \nabla \times \boldsymbol{B} = \mu_0 \left(\boldsymbol{J} + \epsilon_0 \frac{\partial \boldsymbol{E}}{\partial t} \right). \tag{2.2}$$

The electric and magnetic fields can be written in terms of a scalar potential $\phi(\boldsymbol{r}, t)$ and a vector potential $\boldsymbol{A}(\boldsymbol{r}, t)$, according to the relations

$$\boldsymbol{B} = \nabla \times \boldsymbol{A}, \quad \text{and} \quad \boldsymbol{E} = -\nabla \phi - \frac{\partial \boldsymbol{A}}{\mathrm{d}t}. \tag{2.3}$$

There is some ambiguity in the definitions: for any function $f(\boldsymbol{r}, t)$ the *gauge transformations*

$$\boldsymbol{A} \to \boldsymbol{A} + \nabla f, \quad \text{and} \quad \phi \to \phi - \frac{\partial f}{\partial t} \tag{2.4}$$

leave the electric and magnetic fields unchanged. Because of this ambiguity in the potentials, classical electromagnetism treats them as mathematical fictions that are useful, but not of the same physical significance as the fields $\boldsymbol{E}$ and $\boldsymbol{B}$. In quantum theory, the potentials seem to be of more fundamental significance, as indicated by the existence of the well-known Aharonov–Bohm effect [1, 2], and by the role of gauge potentials in relativistic formulations of quantum field theory.

In relativistic formulations, the electric and magnetic fields become mixed with each other under Lorentz transformations. Rather than separate E and B fields, the physically relevant field is the second-rank electromagnetic field tensor,

$$F_{\mu\nu} = \begin{pmatrix} 0 & E_x/c & E_y/c & E_x/c \\ -E_x/c & 0 & -B_z & B_y \\ -E_y/c & B_z & 0 & B_x \\ -E_z/c & -B_y & B_x & 0 \end{pmatrix}, \qquad (2.5)$$

$$= \partial_\mu A_\nu - \partial_\nu A_\mu$$

where μ and ν are space-time indices, running from 0 (for time) to 3 (where 0 to 3 represent space directions). Here, A_μ is a four-vector with components (ϕ, A_x, A_y, A_z), and c is the speed of light in vacuum.

Rather than static electromagnetic fields, our main concern will be electromagnetic waves. In the next section, the chief tool for studying such waves, the Helmholtz equation, will be introduced.

2.2 Helmholtz equation and wave optics

Recall that Maxwell showed the existence of electromagnetic waves that satisfy the *wave equation*,

$$\left(\nabla^2 - \frac{1}{c^2}\frac{\partial^2}{\partial t^2}\right) E_j(\mathbf{r}, t) = 0, \qquad (2.6)$$

where $j = x, y, z$ labels the spatial components, and

$$\nabla^2 = \frac{\partial^2}{\partial x^2} + \frac{\partial^2}{\partial y^2} + \frac{\partial^2}{\partial z^2}$$

is the Laplacian. We will allow the field $E(\mathbf{r}, t)$ to be complex; the actual physical electric field is then given by its real part. Maxwell's wave equation and the related Helmholtz equation (see below) describe the propagation of waves through vacuum at the speed

$$c = \frac{1}{\sqrt{\epsilon_0 \mu_0}} \approx 299\,792\,458 \text{ m s}^{-1} \approx 3 \times 10^8 \text{ m s}^{-1}. \qquad (2.7)$$

Inside matter, the speed is reduced to $v = \frac{c}{n}$, where n is the refractive index of the material.

The archetypal example of an electromagnetic wave is a plane wave. Wavefronts are defined to be surfaces of constant phase. The wavefronts for plane waves are a set of parallel planes of infinite extent. The entire set propagates in the direction perpendicular to surface of each plane. The wave carries momentum $\mathbf{p}$ pointing in the propagation direction. The corresponding Poynting vector $S = \frac{1}{\mu_0} E \times B$, which describes the energy flow, points in this direction as well. The electric and magnetic fields are perpendicular to each other and to the Poynting vector. Although they are simple and serve as a convenient approximation in many circumstances, in real life there are no true plane waves, due to their infinite spatial extent and the infinite energy they carry.

The Maxwell wave equation has solutions describing a variety of different types of waves, in addition to plane wave solutions. For example, the waves from an approximate point source like a light bulb are spherical waves, that propagate outward in all directions. However, here we are concerned with waves that are directional: they propagate as beams along some preferred axis. In the following, we will always take this propagation direction to be the z-axis, so that the electric and magnetic fields of transverse waves will be in the x–y plane. In order to study such beams, it is convenient to go from the general form of the Maxwell wave equation to a special case, called the *Helmholtz equation*. Let ω be the angular frequency of the wave, and then separate the time dependence of the field off from the space dependence: $E(r, t) = E(r)e^{-i\omega t}$. Substituting this into the wave equation, we arrive at the Helmholtz equation,

$$(\nabla^2 + k^2) E_j(r) = 0. \qquad (2.8)$$

Here, the magnitude of the wavevector k is given by the wavenumber $k = \frac{2\pi}{\lambda} = \frac{\omega}{c}$. The Helmholtz equation is the usual starting point for studying any type of directed electromagnetic wave motion.

We will usually be looking only at the region along the beam, close to the axis. In this so-called *paraxial region*, all angles from the beam axis can be assumed to be small. In this case, the three spatial components of the field never mix together and can therefore be treated independently of each other. So henceforth, we will drop the spatial index j from the field components E_j. This means, for example, that the Helmholtz equation will be written as

$$(\nabla^2 + k^2) E(r) = 0. \qquad (2.9)$$

Care should be taken, however, to keep in mind when using expressions like this that E is one of the *components* of the field, and not the *magnitude*.

Let us define some terminology and notation for the following sections. Recall that in quantum mechanics, the momentum and the wavevector are proportional, $p = \hbar k$, so it is common to use the words 'momentum' and 'wavevector' interchangeably. In the following, the beam axis will always be taken to be the z-axis, and we will use cylindrical coordinates (r, z, ϕ) throughout, where ϕ (the *azimuthal* angle) is the angle about the z-axis. The two-dimensional coordinate vector in the x–y plane (the *transverse* plane) will be denoted $r_\perp$. The z-direction will be referred to as the *longitudinal* direction. k_z and k_r are then the longitudinal and transverse wavevectors of k. These components must obey $k_r^2 + k_z^2 = k^2$, where $k = \frac{2\pi}{\lambda}$ and $k_r = \sqrt{k_x^2 + k_y^2}$.

One simple generalization of the Helmholtz equation occurs by allowing the material through which the wave propagates to have a position-dependent index of refraction, $n(\mathbf{r})$. Using the fact that $ck = n\omega$, the k^2 term in equation (2.9) now acts like a spatially-dependent potential energy term. This is a useful fact that allows the properties of a beam inside a material to be tailored in a number of ways in *graded-index (GRIN) materials* [3], where the refractive index is deliberately engineered to vary in a desired manner. For example, GRIN materials allow the design of lenses with flat surfaces. The fact that the k^2 term can be treated as a potential energy term also raises the issue that there is a close formal similarity between the Helmholtz equation and the time-independent Schrödinger equation. Recall that the time-independent Schrödinger equation is given by

$$-\frac{\hbar^2}{2m}\nabla^2 \psi(\mathbf{r}) + V(\mathbf{r})\psi(\mathbf{r}) = E\psi(\mathbf{r}). \tag{2.10}$$

By choosing the origin of the energy scale so that $E = 0$ and by taking $V(\mathbf{r}) = -\frac{\hbar^2}{2m}k^2$, this becomes identical to equation (2.9). This equivalence is often used to simulate quantum problems by means of optical systems. The process can go the other way as well: the Airy beams discussed in chapter 6 were originally derived as solutions to the Schrödinger equation [4].

From a quantum mechanical viewpoint, light waves are formed from a stream of particle-like excitations of the electromagnetic field, called photons. Each photon of frequency ν carries an energy of $E = h\nu = \hbar\omega$ and momentum $p = \frac{E}{c}$. We will return to photons and some quantum aspects of light in chapter 8; until then, classical descriptions will suffice for our purposes.

Our concern will be primarily with the paraxial case, where attention is restricted to the region near the propagation axis. In that case, an argument similar to the separation of variables argument used to obtain the Helmholtz equation from the wave equation may be repeated by defining $E(\mathbf{r}) = u(\mathbf{r})e^{ik_z z}$. This is not *quite* a separation of variables, since u is still a function of z, but we assume that u is much more slowly varying in z than the exponential is. This is guaranteed by remaining in the region where

$$\left|\frac{\partial^2 u}{\partial z^2}\right| \ll \left|k_z \frac{\partial u}{\partial z}\right|. \tag{2.11}$$

Then the Helmholtz equation can be written in paraxial form [5] as

$$\left(\frac{\partial^2}{\partial x^2} + \frac{\partial^2}{\partial y^2}\right)u(\mathbf{r}) + 2ik\frac{\partial u(\mathbf{r})}{\partial z} = 0 \tag{2.12}$$

or

$$\left(\frac{\partial^2}{\partial x^2} + \frac{\partial^2}{\partial y^2}\right)E(\mathbf{r}) + 2ik\frac{\partial E(\mathbf{r})}{\partial z} + 2k^2 E(\mathbf{r}) = 0. \tag{2.13}$$

The remainder of the book will consist of examining specific solutions to the paraxial Helmholtz equations, equations (2.12) or (2.13).

Bibliography

[1] Aharonov Y and Bohm D 1959 Significance of electromagnetic potentials in quantum theory *Phys. Rev.* **115** 485
[2] Peshkin M and Tonomura A 1989 *The Aharonov-Bohm Effect* (Berlin: Springer)
[3] Saleh B E A and Teich M C 2007 *Fundamentals of Photonics* 2nd edn (Hoboken, NJ: Wiley)
[4] Berry M V and Balazs N L 1979 Nonspreading wave packets *Am. J. Phys.* **47** 264
[5] Goodman J W 2004 *Introduction to Fourier Optics* 3rd edn (New York: W H Freeman)

IOP Concise Physics

A Guided Tour of Light Beams
From lasers to optical knots
David S Simon

Chapter 3

Gaussian beams and lasers

3.1 Lasers

One of the chief tools of modern optical science is the laser. As the source of high-quality, high-intensity coherent light beams, its invention led to an explosion of developments in optics, particularly quantum and nonlinear optics, and to a flow of corresponding technological applications that continues to accelerate today. Among other things, the laser provides the light source that allows the creation of all of the specialized light beams to be discussed in the following chapters. Although the beams themselves are the main topic of interest, in this section we briefly cover the essential ideas behind the workings of laser light sources. Lasers can either be pulsed or continuous wave; our concern here will be exclusively with continuous wave lasers.

The first optical laser was built in 1960 by Theodore Maiman, based on prior theoretical work by Charles Townes and Arthur Schawlow. It followed the invention of the maser (the microwave analog of a laser) in 1953 by Maiman, James Gordon, and Herbert Zeiger. The laser provided a reliable source of high intensity coherent light that could be produced with a sharply-peaked spectrum, and provided a valuable research tool that led quickly to advances in a wide range of scientific fields. By the 1980s, lasers could be mass-produced cheaply enough to be used in commercial products like compact disk players.

Laser action is based on the idea of stimulated emission, first discussed by Einstein in 1917. In stimulated emission, the presence of an initial photon of frequency ν and spatial momentum k will stimulate nearby atomic electrons in excited states to emit additional photons into the same quantum state. (The state of the photon or of a beam as a whole is often called a *mode* of the electromagnetic field.) This assumes that the initial frequency is tuned to a resonance of the atom and that there are unoccupied lower-energy states for the electrons to drop into. If there are many electrons in the excited state and many unoccupied holes in the ground state, then the passage of a single resonant photon will cause a sudden avalanche of photon emissions.

A schematic depiction of a typical laser arrangement is shown in figure 3.1. A so-called *gain medium* is contained inside a resonant cavity bounded at the two ends by mirrors, usually spherical in shape. The mirror at one end is partially transmitting to allow the laser light to escape. The resonant cavity is required to control the allowed frequencies, and to guarantee that all the emitted photons are in phase with each other; any light that is out of phase or that has a different wavelength will tend to destructively interfere with itself upon multiple passages through the cavity. The gain medium contains the atoms that will emit the light, once the electrons in those atoms have undergone *population inversion*: energy has to be pumped into the system to elevate electrons from the ground state to a higher energy state. Once there are more electrons in the excited state than the ground state, then the stimulated emission rate by the higher level will be greater than the absorption rate of the ground state electrons. The form of the resonance cavity (primarily its length and the shape of its mirrors) will determine the type of resonant state that arises in the cavity and the type of beam mode that is emitted. The resonant mode in the cavity can become extremely complex; it has even been shown that self-similar fractal modes are possible [1].

The gain medium is typically a gas (such as helium neon, HeNe, or carbon dioxide, CO_2), or a solid. Solid-state lasers include both crystalline materials such as ruby (aluminum oxide with chromium impurities, Al_2O_3:Cr) or semiconductors like gallium arsenide (GaAs). Organic dyes in liquid solutions may also be used, often in frequency-tunable lasers. Instead of a fixed resonant cavity, some lasers are made from optical fiber, where the fiber itself (doped with materials such as erbium and ytterbium ions, and with some of reflective surface at the ends) serves as both the cavity and the active gain medium. The pumping can come in many forms, including optical, electrical, or chemical. For example, a DC current or radio-frequency AC current can be passed through the gas to provide the energy for inversion; this is a common method for gas lasers. Optical pumping using bright bursts of light from a flash tube, arc lamp, or laser diode are common for solid state lasers.

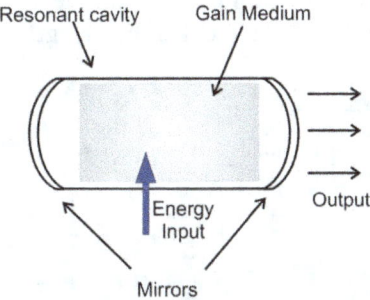

Figure 3.1. The resonant cavity of a laser. Energy input to the gain medium causes population inversion in the material. Light emitted from the excited state undergoes multiple reflections between the mirrors. Destructive interference during these repeated passages causes frequencies far from the resonant frequency to be damped out. The resonant frequency is determined by the distance between the mirrors. The mirror at the right end is partially transmissive, so that the light can eventually escape.

In order to produce laser light, at least three energy levels in the atoms of the gain medium need to be used. A three-level structure (shown in figure 3.2(a)) is the simplest, but a four-level structure (figure 3.2(b)) is more common because of its higher efficiency, so we briefly discuss here the four-level version. The external power input is used to pump electrons from the ground state E_0 to the highest level, E_3. This excited level is of very short lifetime, and decays rapidly by non-radiative means (by thermal phonon generation, without associated photons) to the longer-lived state E_2. The laser light is emitted from E_2, with stimulated emission from E_2 to E_1, leading to an exponentially-growing avalanche of decays. E_1 then decays rapidly by non-radiative means back to the ground state.

The light exiting the laser is notable not only for its high intensity, but also for its narrow bandwidth and high degree of coherence. *Coherence* means that the various waves produced in the material have a fixed phase relationship to each other, so that they can have stable interference patterns, at least over some finite time period (the *coherence time*). Maintaining this fixed relation means that the frequencies must not be too far apart. The presence of the resonant cavity weeds out any light that differs too much from the resonant frequency ν_0. The remaining uncertainty $\Delta \nu$ about this central frequency is determined by the properties of the cavity, and can be made very small compared to ν_0. The coherence time τ_c is then closely related to the bandwidth $\Delta \nu$. Essentially, two photons emitted by the laser at times t and $t + \Delta t$ will interfere with each other if $\Delta t < \tau_c$, but the interference pattern begins to wash out once Δt begins to exceed τ_c. The distance the light travels in one coherence time is often called the longitudinal coherence length, $L_c = c\tau_c$. Coherence times and bandwidths are inversely proportional to each other,

$$\tau_c = \frac{1}{\Delta \nu} = \frac{\lambda^2}{c \Delta \lambda}, \tag{3.1}$$

where the last equality is due to the fact that the wavelength and frequency are related by $\nu = \frac{c}{\lambda}$, and so $\Delta \nu = \frac{c \Delta \lambda}{\lambda^2}$. (A minus sign has been dropped, since τ_c, $\Delta \nu$, and $\Delta \lambda$ are always taken to be positive.) The coherence length of GaAs and HeNe lasers are typically on the order of a millimeter and tens of centimeters, respectively. This should be compared to a typical coherence length of 10^{-6} m (coherence time

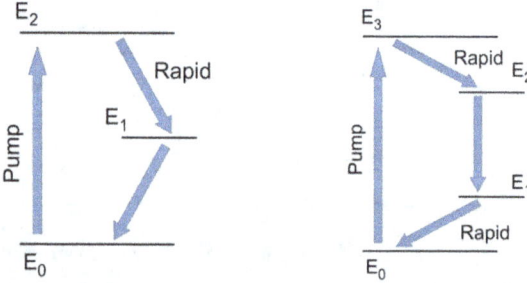

Figure 3.2. Possible energy level arrangements for the gain medium in a laser. (a) A three level system. (b) The more common four-level system. In the four-level version, a large population of electrons is moved into E_2, followed by stimulated emission from E_2 to E_1.

~10^{-14} s) for broadband thermal light (blackbody radiation) emitted by a room temperature object.

The allowed wavelengths depend on both the cavity and the atomic properties of the gain material. Suppose the resonant cavity is of length L and that n is the average refractive index of the gain medium. Since there must be nodes at the mirrors, the allowed wavelengths and angular frequencies in the cavity are related to the cavity length by

$$L = m\frac{\lambda}{2} = \frac{m\pi c}{n\omega}, \quad (3.2)$$

where m is any positive integer. So the resonant frequencies of the cavity are

$$\nu_m = \frac{mc}{2nL}, \quad m = 1, 2, 3\ldots \quad (3.3)$$

The atomic transition that emits the light is not perfectly sharp: the Heisenberg uncertainty relation guarantees that the transition frequency has a finite uncertainty or linewidth (figure 3.3). This linewidth is the width of the spectral line-shape $g(\nu)$ of the atomic state, which gives the amplitude for emitting a photon of frequency ν near the resonant transition frequency of the atom. This line-shape is typically Lorentzian or Gaussian in shape and is broadened by collisions between the atoms, by temperature-induced Doppler shifts, and other effects. Multiple cavity resonance modes may lie within the envelope formed by the atomic line-shape; if they are all retained then the laser is referred to as *multi-mode*; more often a Fabry–Perot etalon (see for example [2]) or some other wavelength-selective device is placed in the cavity to eliminate all but one frequency mode.

The most common output modes for lasers are the so-called transverse-electric (TEM) modes T_{mn}, produced by lasers with spherical mirrors at the ends. These are propagating modes in the z direction with electric and magnetic fields tangent to the x–y plane. The electric field, determined by a pair of integers m and n is given in the transverse x–y plane by

$$E_{mn}(x, y) \sim H_m\left(\frac{\sqrt{2}x}{w}\right) H_n\left(\frac{\sqrt{2}y}{w}\right) e^{-\left(\frac{x^2+y^2}{w^2}\right)} e^{\phi(z)}, \quad (3.4)$$

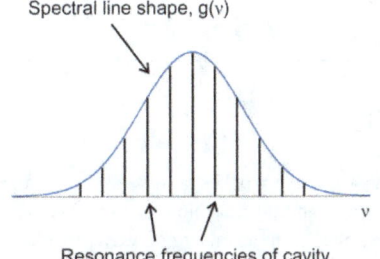

Figure 3.3. The cavity has a sequence of narrow resonant lines, modulated by the much wider spectrally-broadened line width of the atoms.

where E_0 is the field amplitude, $\phi(z)$ is a phase factor, $H_n(x)$ is the Hermite polynomial (appendix A.4), and the parameter w is called the waist size. The lowest order case, $m = n = 0$, is called the *Gaussian mode* and will be discussed in detail in the next section, and the higher order (Hermite–Gauss modes) will be treated in section 7.2. These different modes essentially have wavefronts of the same shape, but the spatial distribution of the amplitude differs.

3.2 Gaussian beams

Before moving on to more exotic types of beams, we first review standard Gaussian or TM_{00} beams. The beam from a laser typically has a Gaussian intensity profile in the plane transverse to propagation. The shape of the beam as it propagates along the z-axis is shown in figure 3.4. The curves shown consist of points at which the intensity is smaller than the on-axis intensity by a factor of e^2. Let w_0 be the beam radius at its narrowest point (the *beam waist*), and choose coordinates so that $z = 0$ is at the location of the waist. The maximum intensity I_0 occurs on-axis at $z = 0$. Further, define the *Rayleigh range* z_0 to be the distance at which the beam cross-section is twice as large as at the waist. The on-axis intensity at the Rayleigh range is therefore half of its value at the waist: $I(z_0) = \frac{1}{2}I_0$. The *depth of focus* (or *confocal parameter*) is the distance $b = 2z_0$ over which the intensity varies by a factor of no more than 2. The Rayleigh range and the waist radius are related by

$$w_0 = \sqrt{\lambda z_0 / \pi}. \tag{3.5}$$

The amplitude E and intensity I of the Gaussian beam can be expressed in the forms

$$E(r, z) = \sqrt{I_0} \frac{w_0}{w(z)} e^{-\frac{r^2}{w^2(z)}} e^{ikz + \frac{ikr^2}{2R(z)} - i\zeta(z)}, \tag{3.6}$$

$$I(r, z) = |E(r, z)|^2 = I_0 \left(\frac{w_0}{w(z)}\right)^2 e^{-\frac{2r^2}{w^2(z)}}. \tag{3.7}$$

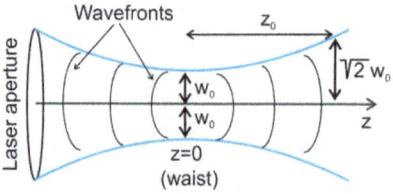

Figure 3.4. A Gaussian beam. The waist, at $z = 0$, has radius w_0. The distance at which the beam has spread to double the area at the waist is the Rayleigh range, z_0. Note that the wavefronts (the surfaces of constant phase) are curved. Approaching the waist, they form a converging wave, then the curvature direction flips after the waist to form a diverging wave. Exactly at the waist, the curvature is infinite, so the wavefront is flat. Therefore, near the waist the laser is an approximate plane wave. The spread of the beam is greatly exaggerated here to make the behavior clear.

Here, the beam radius at distance z from the waist is

$$w(z) = w_0 \left[1 + \left(\frac{z}{z_0}\right)^2\right]^{1/2}, \tag{3.8}$$

and the radius of curvature of the wavefront is

$$R(z) = z\left[1 + \left(\frac{z_0}{z}\right)^2\right]. \tag{3.9}$$

The factor

$$\zeta(z) = \tan^{-1}\left(\frac{z}{z_0}\right) \tag{3.10}$$

is the *Guoy phase*, which varies from $-\frac{\pi}{2}$ to $+\frac{\pi}{2}$ as the beam propagates from $z = -\infty$ to $z = +\infty$. It can verified by direct substitution that a field of the form of equation (3.6) satisfies the Helmholtz equation. The total power carried by the beam is given by $P = \frac{\pi}{2}I_0 w_0^2$.

These beams are initially converging as they leave the laser until they reach a minimum radius (at the waist), and then begin to slowly diverge as a result of diffraction (see figure 3.5). The reason for this contraction and expansion can easily be seen by drawing a few light rays, as in figure 3.5, and noting that the probability of momentum vectors at large angles from the axis drops rapidly with increasing angle. More specifically, the radial components of momentum k_r are normally distributed, with a mean of zero.

Asymptotically, the envelope expands away from the waist region at an angle

$$\theta_0 \approx \frac{\lambda}{\pi w_0}, \tag{3.11}$$

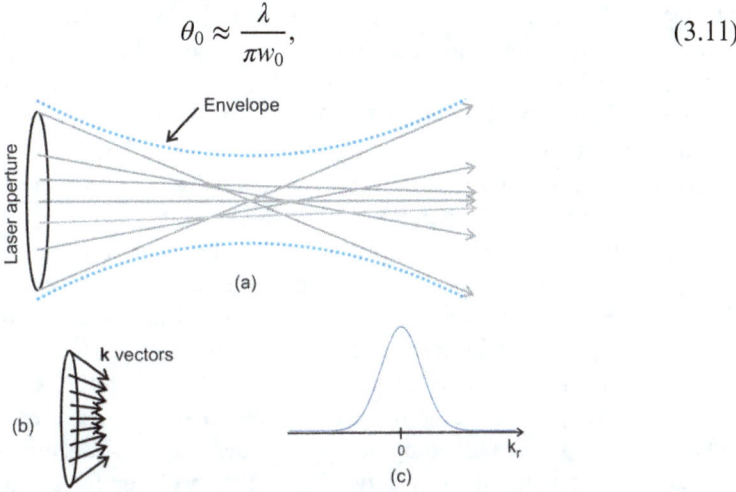

Figure 3.5. (a) Drawing light rays illustrates the form of the beam envelope (dashed curve). The rays are at a range of angles from the z-axis, so that the focus is spread over a finite region (the vicinity of the beam waist) instead of a single point. The bulk of the rays have wavevectors that lie in the interior of a cone (b), with their transverse component k_r mostly at small angles from the axis; more specifically, the transverse momenta follow a Gaussian distribution centered at $k_r = 0$ (c).

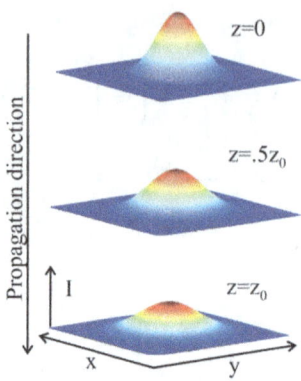

Figure 3.6. The intensity profile of a Gaussian beam at (a) the waist, (b) after $\frac{1}{2}$ of the Rayleigh range, and (c) at one Rayleigh range. The intensity remains normally distributed, but spreads out in the transverse plane as the beam propagates. Since the total power (the integral of the intensity over the transverse plane) must stay constant, the maximum intensity must decrease with distance. The point of maximum intensity in each plane occurs on the axis.

known as the *beam divergence*. Note that the more focused the beam (the smaller the waist size), the more rapidly the beam diverges; this is consistent with equation (3.5) and stems from the Fourier transform relation between the position in the transverse plane and the transverse wavevector (momentum). In order to make the beam better collimated (more beamlike and less divergent), the beam has to be made wider. A cross section of the beam intensity is shown at several points along the propagation axis in figure 3.6. Note that the beam not only expands in size and weakens with distance, as expected, but also that it consists of a maximum at the center that tapers off asymptotically with radius. There are no nodes (points of zero intensity) anywhere. This is in contrast with the beams we will look at later, which will often have a central node on the axis, and several additional nodal rings circling the axis.

The spreading of the Gaussian beam with propagation distance is an inevitable result of diffraction. Whenever a beam of light is localized to a small region (such as by passing it through a hole in a screen, or, as in the current case, by focusing it to a small spot at the waist), each point in the localized region can be thought of emitting Huygens wavelets that expand outward in all directions from that point. After traveling distance r from the emission point, each wave gains a phase factor $e^{-i k \cdot r}$. These wavelets then interfere with each other: the wavelets emitted from nearby points will propagate different distances before reaching the same observation point, and as a result will be out of phase with each other when they combine. Far from the axis, k_r is large, so wavelets that have travelled slightly different distances can have very different phases, leading to mostly destructive interference. Near the axis, $k_r \approx 0$ and the value of $\mathbf{k} \cdot \mathbf{r}$ tends to be much closer to a constant for all of the wavelets at fixed z; the result is that the interference now tends to be constructive. Therefore, away from the axis the interference is predominantly destructive leading to low intensity, while near the axis it is predominantly constructive, leading to the central bright spot. As the beam propagates, the predominantly constructive region expands (since the rays with small k_r have had time to travel farther from the axis), causing the beam to spread. One of the remarkable properties of the Bessel beam that will be discussed in chapter 5 is that, to a good approximation, this diffractive beam spread does *not* occur to a noticeable extent, at least not until a certain minimum distance is reached.

For more detail on Gaussian beams, as well as the principles behind the lasers used to produce them, see [2–4].

3.3 Coherent and squeezed states

Although light is often treated using a classical description, in reality it is (like everything else) actually a quantum mechanical phenomenon. In particular, it must have some quantum uncertainty attached to it: repeated measurements of the same quantity will tend to fluctuate randomly about some mean value. For any given variable η, knowledge of the typical fluctuation size $\Delta \eta$ is often as important for applications as knowledge of the mean, $\bar{\eta}$.

Recall that pairs of variables in quantum mechanics are called *conjugate* if they obey the so-called *canonical commutation relations*; the archetypal examples are the position $\hat{x}$ and momentum $\hat{p}$, which obey

$$[\hat{x}, \hat{p}] = i\hbar, \tag{3.12}$$

where the brackets represent the *commutator*, $[\hat{x}, \hat{p}] = \hat{x}\hat{p} - \hat{p}\hat{x}$. The *Heisenberg uncertainty* puts a lower limit on the products of the uncertainties of such pairs of conjugate variables; for example, the position and momentum uncertainties of an individual electron or photon must obey

$$\Delta x \Delta p \geqslant \frac{1}{4} \left| [\hat{x}, \hat{p}] \right| = \frac{\hbar}{4}. \tag{3.13}$$

Notice that it is the uncertainty *product* that is bounded from below; each individual uncertainty on its own can be made arbitrarily small, but only at the expense of making the other uncertainty larger.

Given this state of affairs, it is natural to ask what kind of light beam achieves the minimal possible uncertainty product? In other words, what is the most classical quantum state of light? The answer is called a *coherent state*.

Coherent states, or minimal uncertainty wave packets, were originally discussed by Schrödinger in 1926 for particles moving in harmonic oscillator potentials, but the study of *optical* coherent states took off with the work of Roy Glauber and E C G Sudarshan in the early 1960s.

Recall that a complex number z requires two real numbers to describe it. z can be described by such a real pair in two different ways. In the polar decomposition, a complex number is written $z = |z|e^{i\theta}$, in terms of the magnitude $|z|$ and the angle from the real axis in the complex plane. But z may also be described by giving its real and imaginary parts, $z = x + iy$. When we describe the electric field along the polarization direction by a complex field amplitude, there is an analogous pair of descriptions. The field can be written in polar form in terms of the real amplitude (the maximum magnitude) and the phase. The other description, in terms of real and imaginary parts, is provided by the two field *quadratures*.

To define the quadratures, first remember that in quantum mechanics, one defines *creation* and *annihilation operator*, $\hat{a}^\dagger$ and $\hat{a}$, which add or remove a photon from a given state. Let the Dirac ket $|\alpha\rangle$ denote the state of an electromagnetic field. Then the state is a coherent state if it is an eigenstate of the annihilation operator,

$$\hat{a}|\alpha\rangle = \alpha|\alpha\rangle. \tag{3.14}$$

The complex number α is the complex amplitude of the state. The photon number operator $\hat{n} = \hat{a}^\dagger \hat{a}$ measures the number of photons in the state, and it has mean value $\langle \hat{n} \rangle = |\alpha|^2$ in a coherent state $|\alpha\rangle$, with uncertainty $\Delta n = \langle n \rangle^{1/2} = |\alpha|$. The uncertainty in the phase of the state is $\Delta \phi = \frac{1}{2|\alpha|}$. $\hat{n}$ and $\hat{\phi}$ provide the analog of the polar-coordinate description for the field.

The two field quadratures are then defined by

$$\hat{X} = \frac{1}{2}(\hat{a}^\dagger + \hat{a}) \tag{3.15}$$

$$\hat{P} = \frac{i}{2}(\hat{a}^\dagger - \hat{a}). \tag{3.16}$$

These have uncertainties $\Delta X = \Delta P = \frac{1}{2}$. Note that the uncertainties are independent of α, so the circular uncertainty area in phase space (the *Wigner distribution*) representing the coherent state (figure 3.7(a)) is the same size for *all* coherent states, independent of amplitude. The increase in $\Delta \phi$ as the amplitude $|\alpha|$ decreases is due solely to the fact that the circle is moving closer to the origin, and therefore subtending a larger angle; it is *not* due to an increase in the actual size of the circular spread.

The quadratures are components of the electric field, but they are analogous in many ways to the mechanical position and momentum variables. In particular, they

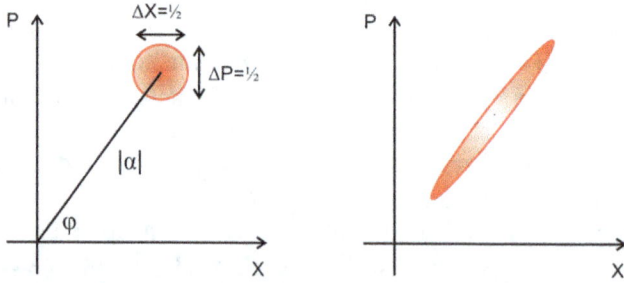

Figure 3.7. Quantum uncertainty spreads the photon quadratures over a finite region of the complex plane (phase space). (a) In the case of a coherent state, this region is circular and of the minimum area allowed by the uncertainty principle. The vacuum state is the same, but centered at the origin. (b) For a squeezed state, the circular region becomes elliptical, but still encloses the same area. In the case shown, the state is squeezed in phase, which increases the uncertainty in $|\alpha|$, or equivalently, in photon number.

obey canonical commutation relations with each other, and for the coherent state they saturate the bounds of the uncertainty relation:

$$\Delta X \Delta P = \frac{1}{4}, \quad (3.17)$$

verifying that the states are minimum-uncertainty states. (Unlike equation (3.12), there is no factor of $\hbar$ in this equation because the quadrature operators $\hat{X}$ and $\hat{P}$ were defined in a way that makes them dimensionless.)

It can also be noted that the vacuum state (the zero-photon state) also has $\Delta X = \Delta P = \frac{1}{2}$. This can be seen either by viewing the vacuum as the $\alpha \to 0$ limit of the coherent state above, or by viewing it as the $n \to 0$ limit of the number state $|n\rangle$; the variances of the number states are given by $\Delta X^2 = \Delta P^2 = \frac{1}{2}\left(\langle n \rangle + \frac{1}{2}\right)$, which goes to the coherent state limit as $\langle n \rangle \to 0$.

Coherent states with different amplitudes form a mathematically complete set, from which any other state can be built. But they are not in general orthogonal to each other. The inner product between two coherent states $|\alpha\rangle$ and $|\beta\rangle$ of different amplitudes α and β is

$$\langle \alpha | \beta \rangle = e^{-\frac{1}{2}(|\alpha|^2 + |\beta|^2) + \alpha^* \beta}. \quad (3.18)$$

A variation of the coherent state is the *squeezed state*. Squeezed states also minimize the uncertainty product, but the uncertainty is not distributed equally among the two quadratures: the circular region in figure 3.7(a) is now elliptical, as in figure 3.7(b). One quadrature is squeezed, leading to smaller uncertainty in that phase space direction. This requires the uncertainty to be increased in the perpendicular direction. Squeezed states are of interest in many areas because they may be used for high-precision measurements. By squeezing the uncertainty in the direction of the variable you wish to measure, that variable can be measured to a greatly enhanced level of accuracy. The cost of course, is that the conjugate variable cannot be measured with much accuracy at the same time. Squeezed states have been used, for example, to make the high precision measurements needed for gravitational wave detection in experiments such as those of the Laser Interferometer Gravitational Wave Observatory (LIGO), where the first verified gravitational wave detection occurred in 2016. Other applications of coherent states include high-precision spectroscopy of weak spectral lines, improved atomic clocks, and continuous-variable quantum cryptography.

3.4 Optical tweezers

The field of optical trapping of atoms, ions, molecules, and other nanoparticles has advanced rapidly in recent decades, and has had enormous consequences across a wide range of fields. One method of doing this was first achieved in 1986 [5]. This is by means of an *optical tweezer*. Optical tweezers are three-dimensional optical traps achieved by using the gradient forces exerted by a beam of light on non-conducting objects.

These forces always push a dielectric particle toward the region of highest optical intensity. Recall that (up to some overall constants), the light intensity is given by the square of the electric field,

$$I = \langle \mathbf{E} \cdot \mathbf{E} \rangle = \langle E^2 \rangle, \tag{3.19}$$

where the brackets denote the time average over the rapidly oscillating field. The force is then proportional to the gradient of the intensity, $\mathbf{F} \sim \nabla I = \langle \nabla |E|^2 \rangle$.

A qualitative picture of how this force arises can easily be seen from figure 3.8. For simplicity, consider a plane wave focused by a lens. The point of maximum intensity will be the focal point of the lens. Now suppose a small dielectric particle (a sphere for example), is placed near the focal point, but displaced slightly away from it. For the sake of specificity, suppose the particle is slightly to the right of the focal point as in figure 3.8(a). When the light hits the sphere from above, it is refracted. The sphere will have a higher index of refraction than the surrounding air or vacuum, so from Snell's law it is easy to see that the light will bend in the direction shown. It will then bend a second time when exiting the sphere. When comparing the ingoing and outgoing rays, it can be seen that the tip of any ray drawn will be rotated clockwise, with increasing component toward the right. Since the light is carrying momentum along its propagation direction, this means the light has gained a net momentum change toward the right. The sphere, therefore, must compensate by gaining an equal and opposite momentum change toward the left in order to conserve momentum. Since force is rate of change of momentum, $\mathbf{F} = \frac{d\mathbf{p}}{dt}$, it follows that the sphere feels a force to the left, or in other words, toward the focal point. Similarly, if the sphere is displaced downward from the focal point, it will feel an upward force, as seen from applying a similar argument to figure 3.8(b).

A more quantitative argument can easily be given to show that the force should be proportional to the intensity gradient. An electric field applied to a dielectric

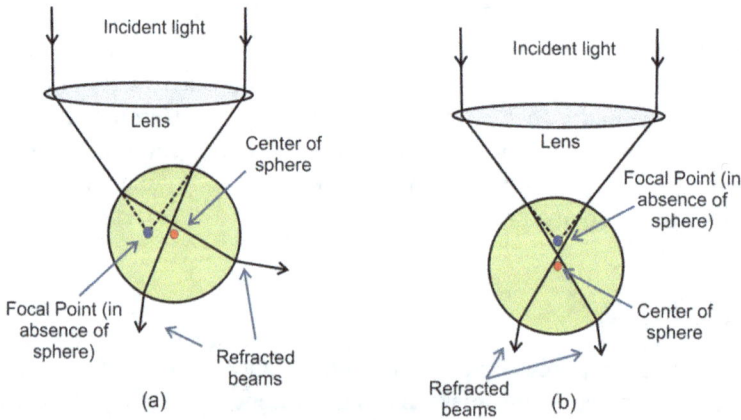

Figure 3.8. The action of light on a small sphere displaced to the right (a) or downward (b) from the point of maximum intensity (the focal point of the lens). Refraction at the surfaces bend the direction of the light rays, causing a change of the optical momentum in the direction of the bending. Momentum conservation then requires that the sphere gain equal and opposite momentum, toward the focal point.

causes a charge separation within molecules, leading to the creation of an electric dipole moment. Recall that for two equal and opposite charges $\pm q$ separated by distance δx, the dipole moment is defined to be $p = q\delta x$. Further recall that the dipole moment induced in the molecule is proportional to the external electric field, $p = \alpha E$, where α is the polarizability of the material.

Suppose the electric field has caused charges $\pm q$ to separate to some pair of points x and $x + \delta x$ separated by a small distance inside the dielectric, and consider the forces on this dipole when it is placed inside a light beam. The beam has both electric and magnetic fields, and these fields will exert a Lorentz force on each charge,

$$F_{\pm} = \pm q\left(E + \frac{dx}{dt} \times B\right). \tag{3.20}$$

The total force on the dipole is given by the sum of the forces on the two individual charges (q at $x + \delta x$, and $-q$ at x), so:

$$F = q\left(E(x + \delta x) - E(x) + \left(\frac{d(x + \delta x)}{dt} - \frac{dx}{dt}\right) \times B\right) \tag{3.21}$$

$$= q\left((E(x) + \nabla E(x) \cdot \delta x) - E(x)\right) + \frac{d(q\delta x)}{dt} \times B + \mathcal{O}(\delta x^2) \tag{3.22}$$

$$= p \cdot \nabla E(x) + \frac{dp}{dt} \times B. \tag{3.23}$$

In the second line the field at $x + \delta x$ was expanded in a Taylor series about x, keeping only terms linear in δx. Using the vector identity

$$E \cdot \nabla E = \frac{1}{2}\nabla(|E|^2) - E \times (\nabla \times E) \tag{3.24}$$

and the Maxwell equation

$$\nabla \times E = -\frac{dB}{dt}, \quad \text{(Faraday's law)} \tag{3.25}$$

we may write

$$E \cdot \nabla E = \frac{1}{2}\nabla(|E|^2) + E \times \frac{dB}{dt}. \tag{3.26}$$

It is then found that equation (3.23) can be rewritten as a gradient plus a term proportional to the Poynting vector, $S = \frac{1}{\mu_0}E \times B$:

$$F = \frac{\alpha}{2}\nabla(|E|^2) + \alpha E \times \frac{dB}{dt} + \frac{dp}{dt} \times B \tag{3.27}$$

$$= \frac{\alpha}{2}\nabla(|E|^2) + \alpha\frac{d}{dt}(E \times B). \tag{3.28}$$

At any fixed point in the beams, the time averaged Poynting vector is constant, so when the average is taken the last term vanishes, leaving the result:

$$\langle F \rangle = \frac{1}{2}\alpha\nabla\langle |E|^2 \rangle = \frac{\alpha}{2}\nabla I. \tag{3.29}$$

The polarizability can also be computed [6], so this result may be written more explicitly as

$$\langle F \rangle = \frac{2\pi n_m a^3}{c}\left(\frac{r^2-1}{r^2+2}\right), \tag{3.30}$$

where a is the radius of the sphere and $r = n_s/n_m$ is the ratio of refractive indices of the sphere and the surrounding medium.

By moving the light source or the lens, the location of the beam focus can be changed. As the maximum intensity point moves in space, the trapped particle will follow. In this way, molecules and other microscopic particles can be moved individually in order to, for example, build specially-tailored nanostructures or to alter chromosomes at the level of individual base pairs.

Upon their invention, optical tweezers immediately found applications in biology. Early applications included the rotation and tethering of bacteria to microscope slides by their flagella [7] and the isolation and manipulation of individual viruses or cells [8–11]. Optical tweezers have been used as components of many different types of microscopes and force probes [12–14]. They have been used for measuring the forces exerted between individual myosin molecules and an actin filament [15], measuring the amount of stretching of individual DNA molecules [16], to study protein folding [17], and for particle sorting [18], among many other applications. By adding an additional laser beam, optical scissors can be created for doing surgery on individual chromosomes and other molecular-scale objects [19].

A further twist will be given (literally) to optical tweezers in section 4.3, where the use of beams with orbital angular momentum will turn them into optical wrenches or spanners. More detailed reviews of optical tweezers and optical trapping can be found in [20–22].

Bibliography

[1] Karman G P, McDonald G S, New G H C and Woerdman J P 1999 Laser optics: Fractal modes in unstable resonators *Nature* **402** 138
[2] Saleh B E A and Teich M C 2007 *Fundamentals of Photonics* 2nd edn (Hoboken, NJ: Wiley)
[3] Siegman A E 1986 *Lasers* (Mill Valley, CA: University Science Books)
[4] Svelto O 1989 *Principles of Lasers* 3rd edn (New York: Plenum)
[5] Ashkin A, Dziedzic J M, Bjorkholm J E and Chu S 1986 Observation of a single-beam gradient force optical trap for dielectric particles *Opt. Lett.* **11** 288
[6] Harada Y and Asakura T 1996 Radiation Forces on a dielectric sphere in the Rayleigh scattering regime *Optics Commun.* **124** 529
[7] Block S M, Blair D F and Berg H C 1989 Compliance of bacterial flagella measured with optical tweezers *Nature* **338** 514

[8] Ashkin A and Dziedzic J M 1987 Optical trapping and manipulation of viruses and bacteria *Science* **235** 1517
[9] Ashkin A, Dziedzic J M and Yamane T 1987 Optical trapping and manipulation of single cells using infrared laser beams *Nature* **330** 769
[10] Ashkin A and Dziedzic J M 1989 Internal cell manipulation using infrared laser traps *Proc. Nat. Acad. Sci. USA* **86** 7914
[11] Ashkin A, Schutze K, Dziedzic J M, Euteneuer U and Schliwa M 1990 Force generation of organelle transport measured *in vivo* by an infrared laser trap *Nature* **348** 346
[12] Liu Y, Sonek G J, Liang H, Berns M W, Konig K and Tromberg B J 1995 Two-photon fluorescence excitation in continuous-wave infrared optical tweezers *Opt. Lett.* **20** 2246
[13] Visscher K and Brakenhoff G J 1991 Single beam optical trapping integrated in a confocal microscope for biological applications *Cytometry* **12** 486
[14] Ghislain L P and Webb W W 1993 Scanning-force microscope based on an optical trap *Opt. Lett.* **18** 1678
[15] Finer J T, Simmons R M and Spudich J A 1994 Single myosin molecule mechanics: piconewton forces and nanometre steps *Nature* **368** 113
[16] Wang M D, Yin H, Landick R, Gelles J and Block S M 1997 Stretching DNA with optical tweezers *Biophys. J.* **72** 1335
[17] Cecconi C, Shank E A, Bustamante C and Marqusee S 2005 Direct observation of the three-state folding of a single protein molecule *Science* **309** 2057
[18] MacDonald M, Spalding G and Dholakia K 2003 Microfluidic sorting in an optical lattice *Nature* **426** 421
[19] Liang H, Wright W H, Rieder C L, Salmon E D, Profeta G, Andrews J, Liu Y, Sonek G J and Berns M W 1994 Directed movement of chromosome arms and fragments in mitotic newt lung cells using optical scissors and optical tweezers *Expt. Cell Res.* **213** 308
[20] Molloy J E and Padgett M J 2002 Lights, action: Optical tweezers *Cont. Phys.* **43** 241
[21] Woerdemann M, Alpmann C, Esseling M and Denz C 2013 Advanced optical trapping by complex beam shaping *Laser Phot. Rev.* **7** 839
[22] Stevenson D, Gunn-Moore F and Dholakia K 2010 Light forces the pace: optical manipulation for biophotonics *J. Biomed. Opt.* **15** 41503

IOP Concise Physics

A Guided Tour of Light Beams
From lasers to optical knots
David S Simon

Chapter 4

Orbital angular momentum and Laguerre–Gauss beams

Orbital angular momentum (OAM) is a concept that is familiar to every physics student, but its importance in optics was not recognized until the 1990s. Since then, the literature on optical OAM has grown exponentially, as researchers have found applications ranging from super-resolution microscopy to fabrication of nanomaterials. Quantum-mechanically entangled pairs of photons with anticorrelated OAM values have also become a valuable tool in high capacity quantum cryptography and other areas.

In this chapter, the idea of optical OAM is introduced. The focus here will be on Laguerre–Gauss beams, which carry quantized values of OAM and which are particularly simple to make. In later chapters, it will seen that other types of light beams, such as the Bessel beam, can also carry non-trivial OAM.

4.1 Polarization and angular momentum in optics

First, recall the idea of orbital angular momentum, as it is introduced in the context of mechanics. Imagine an object of mass m orbiting a fixed point with constant speed v and orbital radius r (figure 4.1(a)). The angular momentum about the orbital axis (the z-axis) is $L = mvr = rp$, where p is the magnitude of the linear momentum. More generally, an object undergoing any type of motion in space (not necessarily an orbital motion) has an angular momentum vector relative to any point P, given by

$$\mathbf{L} = \mathbf{r} \times \mathbf{p}, \qquad (4.1)$$

where $\mathbf{r}$ is the vector pointing from P to the object (figure 4.1(b)). In particular, notice that even straight line motion has angular momentum about P, as long as the particle's path does not collide with P.

In quantum mechanics, angular momentum, like other dynamical variables, becomes an operator that acts on wave functions. The operator form in the position

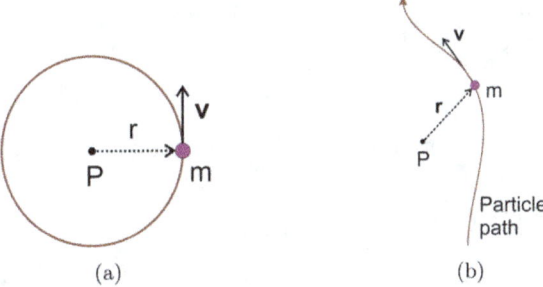

Figure 4.1. (a) A particle undergoing circular motion about a fixed point has orbital angular momentum $L = r \times p$, where $p = mv$ is the linear momentum. (b) A more general motion will still have angular momentum about P, but now L is no longer constant.

representation is obtained from the classical expression simply by replacing the momentum variable with its operator form, $p \to \hat{p} = -i\hbar \nabla$, so that

$$\hat{L} = -i\hbar \,\hat{r} \times \nabla. \tag{4.2}$$

Once a measurement axis is chosen (again usually called the z-axis), the magnitude squared $\hat{L}^2 = \hat{L}_x^2 + \hat{L}_y^2 + \hat{L}_z^2$ of the orbital angular momentum and its z-component $\hat{L}_z$ can be simultaneously diagonalized, so that if the wave function ψ is an angular momentum eigenstate, both of these operators have quantized eigenvalues:

$$\hat{L}^2 \psi(r) = L(L-1)\hbar^2 \psi(r), \qquad \hat{L}_z \psi(r) = m_l \hbar \, \psi(r), \tag{4.3}$$

where $L = 0, 1, 2, 3\ldots$ and $m_l = -L, \ldots, 0, \ldots L$. L and m_l are examples of quantum numbers.

The wavefunction represents some particle, and this orbital angular momentum $\hat{L}$ is due to the motion of that particle; for example, it could be the orbital motion of an electron around an atomic nucleus. However, it turns out that even a particle at rest can have angular momentum. This angular momentum, which is intrinsic to the particle and independent of its state of motion, is called *spin*, and is represented by an operator $\hat{S}$. Its squared magnitude

$$\hat{S}^2 \psi(r) = s(s-1)\hbar^2 \psi(r) \tag{4.4}$$

is constant. The value of s is always the same for a given type of particle: $s = 0$ for spinless particles like the Higgs boson, $s = \frac{1}{2}$ for electrons and protons, and $s = 1$ for photons. More generally, any fermion (particles that obey the Pauli exclusion principle) will have spin that is an odd multiple of $\frac{1}{2}$ (in units of $\hbar$), and any boson (a particle not obeying the exclusion principle) has integer spin. The component along any measurement axis is also quantized:

$$\hat{S}_z \psi(r) = m_s \hbar \, \psi(r), \tag{4.5}$$

where $m_s = -s, ..., 0, ... s$. The total angular momentum of a quantum state is then the sum of the spin and angular momenta:

$$\hat{J} = \hat{L} + \hat{S}. \tag{4.6}$$

The component of angular momentum along the direction of motion of a particle is called *helicity*,

$$\hat{h} = \frac{\hat{J} \cdot \hat{p}}{|p|}. \tag{4.7}$$

Since the cross-product $r \times p$ of equation (4.1) is always perpendicular to the momentum direction, this means the helicity is due entirely to the spin, not to orbital angular momentum. Because light is made of photons, it has spin $s = 1$. Consequently, its possible helicity values should be $m_s = -1, 0, +1$, in units of $\hbar$. However, the massless nature of the photon and the requirement of gauge invariance forces the $m_s = 0$ component to vanish, at least for those photons of interest here, i.e. those that can propagate in light beams. (The vanishing of the $m_s = 0$ component does not apply to the non-propagating cloud of *virtual* photons that make up an electrostatic Coulomb field.) The reason for this would take us too far afield to discuss here; see any book on quantum electrodynamics or quantum field theory such as [1–3] for a discussion. But the point is that light therefore has two possible spin states, like a fermion, despite the fact that it is made of bosons.

Recall that the polarization state of light corresponds to the direction of its electric field: for linear polarization, the field is pointing in a fixed direction, while for circular polarization the field is rotating as the light propagates. The two values $m_s = \pm 1$ correspond to the two possible rotation directions (clockwise or counter-clockwise), representing right-handed or left-handed circular polarization states. Represent the two circular polarization states by $|R\rangle$ and $|L\rangle$. Linear polarizations in the vertical and horizontal direction are then equal superpositions of these two circular polarizations:

$$|H\rangle \equiv \frac{1}{\sqrt{2}}\big(|L\rangle + |R\rangle\big), \quad |V\rangle \equiv \frac{i}{\sqrt{2}}\big(|L\rangle - |R\rangle\big). \tag{4.8}$$

Adding additional phase shifts between the two states in these sums leads to linear polarizations in other directions. Inverting, one may write the circular polarizations in terms of the linearly polarized states:

$$|R\rangle \equiv \frac{1}{\sqrt{2}}\big(|H\rangle + i|V\rangle\big), \quad |L\rangle \equiv \frac{1}{\sqrt{2}}\big(|H\rangle - i|V\rangle\big). \tag{4.9}$$

The existence of polarized light has been known since the 19th century and it was realized early in the history of quantum mechanics that circular polarization was a result of the photon's intrinsic spin. It also quickly became clear that this angular momentum can be transferred to matter [4–7].

It was only realized much later that, in addition to the intrinsic spin angular momentum $\hat{S}$, a photon can also carry orbital angular momentum (OAM) $\hat{L}$ about its propagation axis. This OAM is due to the possibility of the photon state having non-trivial spatial structure in the transverse direction, and was first explored in detail in [8]. There are several excellent reviews of the subject, including [9–11], that the reader may consult for additional detail and applications beyond those discussed here.

Suppose an approximate plane wave is multiplied by an azimuthally-dependent phase shift of the form $e^{il\phi}$, where ϕ is the angle about the propagation axis (figure 4.2). The reader may easily verify that the angular momentum operator $\hat{L}_z = -i\frac{\partial}{\partial \phi}$ then has eigenvalue $L_z = l\hbar$. (From here on, we conform to the standard notation in this field and denote m_l simply as l.) The single-valuedness of the field under $\phi \to \phi + 2\pi$ will force the *topological charge l* to be quantized: $e^{il\phi} = e^{il(\phi+2\pi)}$ can only be true if l is an integer. The phase factor has the effect of tilting the wavefronts by an increasing amount as the axis is circumnavigated, so that the wavefronts have a corkscrew shape, as in figure 4.3. The Poynting vector $S = E \times H$, which gives the magnitude and direction of the energy flow, is always perpendicular to the wavefront. As a result, S is tilted away from the propagation axis by some fixed angle and will rotate about the axis as the wave propagates, providing the non-zero orbital angular momentum carried by the wave.

The topological charge is a special case of the concept of winding number that occurs in topology. Winding numbers count the number of times a path through some space wraps around a hole in that space. Think for example of drawing closed loops on the surface of a torus, and you will quickly see that some of the loops wrap around the donut hole; these cannot continuously be contracted to a point without

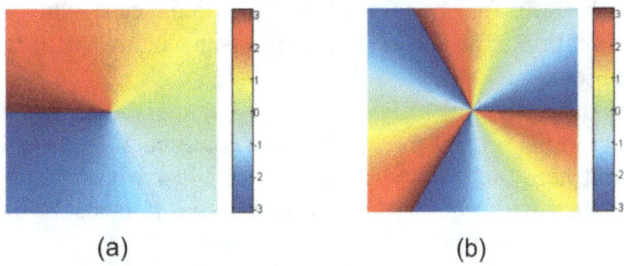

(a)   (b)

Figure 4.2. The azimuthal phase function $e^{il\phi}$, where ϕ is the angle about the z axis. A state with topological charge l has a phase that varies by a total amount $2\pi l$ as the angle goes from 0 to 2π. The figure on the left shows the phase modulo 2π for $l = 1$, while the right shows the same for $l = 3$.

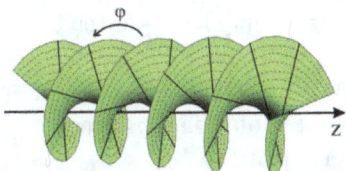

Figure 4.3. Twisted light: an optical wavefront with non-zero orbital angular momentum. Surfaces of constant phase are corkscrew-shaped. The Poynting vector, S must be everywhere perpendicular to the wavefront, so that it rotates as the wave propagates along the z-axis.

leaving the surface of the torus, and so are topologically distinct from contractible circles (those with winding number zero). Paths that wind different numbers of times around the hole form different topological classes. In the case of beams with topological charge $l \neq 0$, the 'hole' in the space of states corresponds to the axis, where the intensity is forced to be zero by a phase singularity. If $l \neq 0$ then contracting any path around the axis would make the phase multiply valued when the radius reaches zero; this is not allowed, so the axis forms a hole where none of the photon amplitude is allowed to exist. The topological charge then counts how many times the phase factor $e^{il\phi}$ circles the origin in the complex plane as a closed path going from $\phi = 0$ to $\phi = 2\pi$ is traced out around the axis.

Closed curves in a given space can be grouped into a set of equivalence classes, with two curves being considered equivalent if they can be continuously deformed into each other. These equivalence classes then form a mathematical group, called the homotopy group [12, 13]. The classification of spaces by their homotopy groups (or by related structures like homology and cohomology groups) is a branch of algebraic topology that plays a large role in many areas of physics, including solid state physics and quantum field theory. The homotopy group defined by the set of phase factors $e^{il\phi}$ is simply the group $\mathbb{Z}$ of integers under addition. The integer topological charge or winding number, l, labels the different equivalence classes in the group. Each equivalence class consists of the set of closed curves on which the phase circles the origin in the complex plane l, times, when the angle in real space circles the beam axis once.

For electromagnetic fields, the separation of angular momentum into spin and orbital angular momentum has complications in the general case [14–20]. But it is simple and unambiguous for optical fields in the paraxial case, where only the region close to the propagation axis need be considered. We will always restrict ourselves to this case when discussing OAM.

A number of different optical beam modes can carry OAM, including higher-order Bessel or Hermite–Gauss modes. We will come back to those other beams in later chapters, but here the focus will be on the Laguerre–Gauss (LG) modes. The field of the LG beam with OAM $l\hbar$ and with p radial nodes is [21]

$$E_{lp}(r, z, \phi) = \frac{E_0}{w(z)} \left(\frac{\sqrt{2}\,r}{w(z)}\right)^{|l|} e^{-r^2/w^2(r)} L_p^{|l|}\left(\frac{2r^2}{w^2(r)}\right) \quad (4.10)$$

$$\times\, e^{-ikr^2z/(2(z^2+z_R^2))} e^{-i\phi l + i(2p+|l|+1)\arctan(z/z_R)},$$

where E_0 is a constant and $w(z) = w_0\sqrt{1 + \frac{z}{z_R}}$ is the beam radius at distance z. $L_p^\alpha(x)$ are the associated Laguerre polynomials [22], and $z_R = \frac{\pi w_0^2}{\lambda}$ is the Rayleigh range and the arctangent term is the Gouy phase. Henceforth, r represents the distance from the axis in the transverse plane.

The index p characterizes the radial structure of the mode: in addition to the central dark spot, u_{lp} has p dark nodal rings. States with different p values can be related by a type of raising and lowering operator and can be shown to be eigenstates

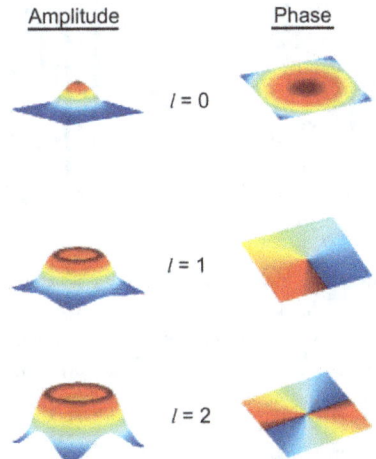

Figure 4.4. Plots of intensity and phase in the transverse plane for several LG modes with different l values. $p = 0$ in each case. For $l = 0$, the phase is rotationally symmetric about the axis, but for $l \neq 0$, it increases linearly with angle as the axis is circled. Note that the amplitude vanishes on the axis for the $l \neq 0$ cases.

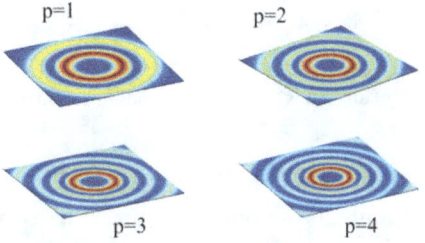

Figure 4.5. Plots of intensity for several LG modes of different p in the transverse plane. $l = 4$ in each case. Notice that in addition to the dark central spot on the axis, the number of nodal rings on which the amplitude vanishes (the dark blue rings) is given by the value of p.

of a differential operator [23–25]. Examples of the intensity and phase of these beams are plotted in figure 4.4 at $p = 0$ for different values of l. Plots of intensity for different p at fixed l are shown in figure 4.5.

4.2 Generation and detection of Laguerre–Gauss beams

There are a number of ways to generate optical Laguerre–Gauss states. One way is by means of *spiral phase plates* (figure 4.6(a)). These are plates whose optical thickness varies azimuthally according to $\Delta z = \frac{l\phi}{k(n-1)}$ [26]. The phase shift gained relative to vacuum propagation at angle ϕ in the material is $l\phi$.

Another common production method is by the use of forked diffraction gratings [27] (figure 4.6(b)); the illuminated spot produced in the first-order diffraction of a Gaussian mode forms an OAM Laguerre–Gauss mode. The value of l is determined by the structure of the dislocation at the center of the grating. Such gratings can be in the form of computer generated holograms or they can be programmed onto the

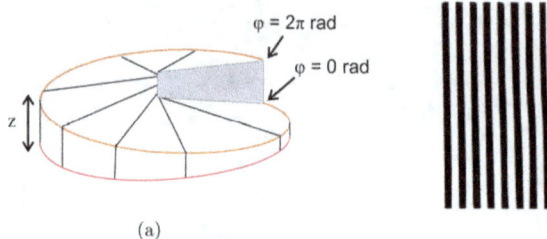

Figure 4.6. Ways to generate OAM states include spiral wave plates (a) and forked diffraction gratings (b). In (a), the spiral wave plate has thickness that increases linearly with azimuthal angle ϕ, so that a phase factor of $e^{il\phi}$ accumulates as the measurement point rotates through angle ϕ about the axis. If the material has refractive index n, then the thickness at angle ϕ is $\Delta z = \frac{l\phi}{k(n-1)}$. In the forked grating of (b), the diffraction pattern has intensity maxima with OAM values determined by the form of the dislocation at the center.

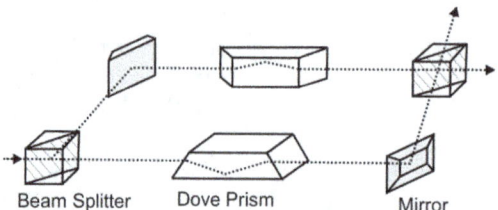

Figure 4.7. An interferometric arrangement for sorting OAM values [28]. The two Dove prisms rotate the beam, and for appropriate choice of angle this arrangement sends even values of OAM to one output port and odd values to the other.

surface of a spatial light modulator (SLM). (SLMs will be discussed in chapter 9.) The grating has a discontinuity at the branch point, where one line splits into several. Note that a common feature of both of these methods is the existence of a singular point in the production plane, at the center of the grating or at the center of the spiral phase plate.

Measurement of the OAM content of a beam can be accomplished several ways. One common method is to use an interferometric arrangement that sorts different l values into different outgoing spatial directions [28–30], so that each OAM component reaches a different detector. An example of an interferometric sorter is shown in figure 4.7 [28]. The basic unit consists of a Mach–Zehnder interferometer with a Dove prism in each arm. The prisms rotate the beams, so that two versions of the beam will interfere with each other either constructively or destructively in the interferometer, depending on the amount of rotation and the value of the OAM. For example, if the OAM is even and the relative rotation angle between the two beams is $2\pi/m$ for even integer m, then destructive interference will occur at one of the two final outputs and constructive at the other. The result is that even and odd values of l are separated into different output ports of the final beam splitter. By combining several layers of such units with progressively smaller rotation angles and introducing appropriate OAM shifts Δl into one outgoing beam of each layer, all values of l can eventually be sorted out. The first stage separates even and odd l (in other words separates values mod 2), the second stage separates values that differ mod 4, the third separates values differing by mod 8, and so on. A similar strategy, with the addition of quarter-wave plates instead of Dove prisms, allows the values of both OAM and spin (polarization) to be sorted simultaneously [29]. All of these sortings can be done even at the single photon level.

A number of other possible methods exist for sorting OAM values, including the use of q-plates [31], polarizing Sagnac interferometers [32], pinhole arrangements followed by Fourier-transforming lenses [33], specialized refractive elements [34–36], and a simultaneous sorter for spin and orbital angular momentum that uses the Pancharatnam–Berry phase has been proposed [37].

4.3 Optical spanners and micropumps

In section 3.4, the idea of optical tweezers was discussed. The use of beams with OAM adds additional capabilities for optical manipulation of microparticles. Annular beams that carry non-zero OAM can be used in optical tweezer setups. These have been shown to confine particles inside the dark central region of the beam and to induce rotation by transferring angular momentum to the particles [38, 39]. It was also found that rod-like bacteria are rotated by the gradient forces so that they align themselves along the beam axis [40], and that high-order Hermite–Gauss modes (section 7.2) could be used to trap and rotate an individual red blood cell [41]. This ability to transfer OAM from the beam to material particles allows rotation of microscopic objects by controlled amounts and at controlled rates, so that the beam serves as an *optical spanner* or wrench.

Building on the same idea, arrays of OAM vortices can also be used to construct reconfigurable microfluidic pumps [42]. The OAM-bearing light beams incident on the fluid transfer angular momentum to dielectric particles in the fluid. These particles begin to rotate around the bright ring of the beam, dragging a layer of fluid with them. Two rows of such vortices, rotating in opposite directions, act as a pump dragging the fluid and any particles floating in it through the space between the two rows (figure 4.8).

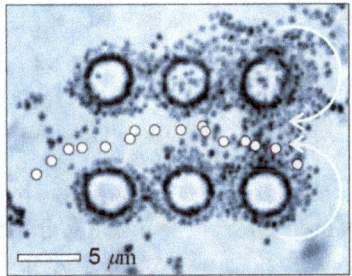

Figure 4.8. A cross-section of an array of six optical vortices passing through a fluid trapped between a microscope slide and a cover slip. The bottom three vortices rotate counter-clockwise, while the top three rotate clockwise. The result is that particles tend to be pumped toward the left through the region between the two rows. The white circles give the trajectory of a single silica sphere over a seven second period, during which the sphere reaches a maximum speed of 5 μm s^{-1}. (Figure reproduced from [42].)

4.4 Harnessing OAM for measurement

Beams with non-zero orbital angular momentum can be used for a variety of optical measurements. For example, rotation rates can be measured with the help of optical OAM [43–47]. Imagine two photons of opposite OAM values, $\pm l$, reflecting off the surface of a rotating object. Let ω be the photon frequency and Ω be the rotational frequency of the object. The two reflected photons experience equal and opposite angular Doppler shifts upon reflection: $\omega \to \omega \pm l\Omega$. Bringing the two reflected photons together leads to an interference pattern with a beat frequency of $2l\Omega$:

$$\text{Intensity} \sim \left| e^{i(\omega+l\Omega)t} + e^{i(\omega-l\Omega)t} \right|^2 = 2(1 + \cos(2l\Omega t)). \tag{4.11}$$

For sufficiently large l, this then allows precise measurements of very small Ω values.

In [48], it has been also been shown that angular displacements (rotation angles) can be measured to high precision using OAM beams. The method makes use of a pair of entangled photons with equal and opposite OAM values. The measurement is done by passing the photons through a specialized type of interferometer and then measuring interference in the coincidence rate (the rate at which both photons are measured simultaneously in two different detectors). This coincidence counting of entangled-photon pairs makes the method inherently quantum mechanical, and allows rotation angles to be measured much more precisely than can be done with classical approaches. Entanglement and coincidence counting will be discussed further in chapter 8.

In microscopy and other imaging methods, optical OAM-based methods have been used to increase both the resolution and contrast when imaging phase objects. A phase object is an object that gives spatially-varying phase shifts to light passing through it at different points, without necessarily diminishing the light intensity being transmitted. These methods arrange for light with non-zero OAM to constructively interfere in regions where the phase is constant, but to destructively interfere in regions of abrupt phase changes (or vice versa). This leads to greatly enhanced contrast at edges of an object, where the phase changes suddenly. See [49–53] for more details and applications of the method.

One further type of measurement can be mentioned. When light of a given OAM value passes through or reflects off of an object, the interaction with the object can cause the OAM to be altered. Not only may the object block some values of l, but it can create new ones due to diffractive effects. As a result, the incoming OAM spectrum (before the object) and the outgoing spectrum afterward may not be the same. The change in OAM spectrum will be determined by the properties of the object and will serve as a sort of fingerprint for the object in the space of OAM values. As a result, the measurements of the ingoing and outgoing OAM spectra can be used to identify the object, or possibly even to reconstruct its shape. This can be done simply by sorting OAM values; no cameras or other spatially-resolving detectors need to be used. Such methods have been studied with both individual photons and with entangled photon pairs [54–58]. One advantage of this approach is that the OAM spectrum is extremely sensitive to rotational symmetries, so that the

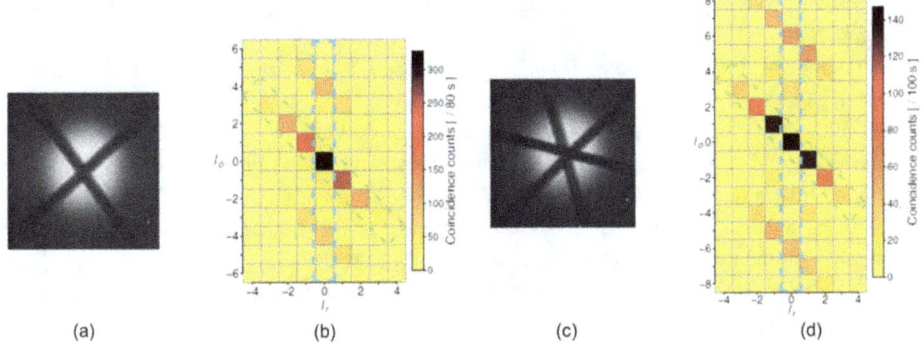

Figure 4.9. Passage of light through an object alters its OAM spectrum. Due to conservation of angular momentum, two entangled photons should always lie on the diagonal when the two OAM values measured in coincidence are plotted. But when an object is placed in one of the two beam paths, the OAM spectrum is altered so that non-diagonal terms appear. An object with four-fold rotational symmetry (a) shifts some of the amplitudes up and down by multiples of four (b) on the plot. Similarly, a six-fold rotational symmetry (c) shifts some of the amplitudes up and down by multiples of six (d). The two axes in (b) and (d) give the OAM, l_o and l_r, of the photons measured in the two detectors (the object and reference detectors). The presence of rotational symmetries can therefore be read directly off the plot. (Figures reproduced from [57].)

presence of these symmetries can be read directly off the coincidence data in the entangled case, as seen in figure 4.9. Further, the OAM spectrum is insensitive to rotations of the object, so that the same object oriented at different angles will give the same signature, and rapidly rotating objects can be identified by the same method as stationary ones.

Bibliography

[1] Zee A 2010 *Quantum Field Theory in a Nutshell* 2nd edn (Princeton, NJ: Princeton University Press)
[2] Ryder L H 1996 *Quantum Field Theory* (Cambridge: Cambridge University Press)
[3] Itzykson C and Zuber J-B 1980 *Quantum Field Theory* (New York: McGraw-Hill, reprinted by Dover Books, Mineola) 2005
[4] Poynting J H 1909 The wave motion of a revolving shaft, and a suggestion as to the angular momentum in a beam of circularly polarised light *Proc. R. Soc. Lond.* A **82** 560
[5] Beth R A 1935 Direct Detection of the angular momentum of light *Phys. Rev.* **48** 471
[6] Beth R A 1936 Mechanical detection and nmeasurement of the angular momentum of light *Phys. Rev.* **50** 115
[7] Carrara N 1949 Coppia e momento angolare della radiazione *Nuovo Cimento* **6** 50
[8] Allen L, Beijersbergen M W, Spreeuw R J C and Woerdman J P 1992 Orbital angular-momentum of light and the transformation of Laguerre–Gaussian laser modes *Phys. Rev.* A **45** 8185
[9] Yao A M and Padgett M J 2011 Orbital angular momentum: origins, behavior and applications *Adv. Opt. Phot.* **3** 161
[10] Torres J P and Torner L (ed) 2011 *Twisted Photons: Applications of Light with Orbital Angular Momentum* (Hoboken, NJ: Wiley)

[11] Franke-Arnold S, Allen L and Padgett M 2008 Advances in optical angular momentum *Laser Photon. Rev.* **2** 299
[12] Nakahara M 2003 *Geometry, Topology and Physics* 2nd ed (Boca Raton, FL: Taylor and Francis)
[13] Nash C and Sen S 1983 *Topology and Geometry for Physicists* (London: Academic)
[14] Barnett S and Allen L 1994 Orbital angular momentum and nonparaxial light beams *Opt. Comm.* **110** 670
[15] van Enk S J and Nienhuis G 1994 Commutation rules and eigenvalues of spin and orbital angular momentum of radiation fields *J. Mod. Opt.* **41** 963
[16] van Enk S J and Nienhuis G 1994 Spin and orbital angular momentum of photons *Europhys. Lett.* **25** 497
[17] Barnett S M 2002 Optical angular-momentum flux *J. Opt. B: Quant. Semiclass. Opt.* **4** S7
[18] Zhao Y, Edgar J S, Jeffries G D M, McGloin D and Chiu D T 2007 Spin-to-orbital angular momentum conversion in a strongly focused optical beam *Phys. Rev. Lett.* **99** 073901
[19] Nieminen T A, Stilgoe A B, Hechenberg N R and Rubinstein-Dunlop H 2008 Angular momentum of a strongly focused Gaussian beam *J. Opt. A: Pure Appl. Opt.* **10** 115005
[20] Santamoto E 2004 Photon orbital angular momentum: problems and perspectives *Fortschr. Phys.* **52** 1141
[21] Allen L, Padgett M and Babiker M 1999 The orbital angular momentum of light *Prog. Opt.* **39** 291
[22] Arfken G, Weber H and Harris F E 2012 *Mathematical Methods for Physicists: A Comprehensive Guide* 7th edn (London: Academic)
[23] Karimi E and Santamoto E 2012 Radial coherent and intelligent states of paraxial wave equation *Opt. Lett.* **37** 2484
[24] Karimi E, Boyd R W, de Guise H, Řeháček J, Hradil Z, Aiello A, Leuchs G and Sánchez-Soto L L 2014 Radial quantum number of Laguerre-Gauss modes *Phys. Rev.* A **89** 063813
[25] Plick W N, Lapkiewicz R, Ramelow S and Zeilinger A 2013 The forgotten quantum number: a short note on the radial modes of Laguerre-Gauss beams arXiv:1306.6517 [quant-ph]
[26] Beijersbergen M W, Coerwinkel R, Kristensen M and Woerdman J P 1994 Helical-wavefront laser beams produced with a spiral phaseplate *Opt. Commun.* **112** 321
[27] Yu Bazhenov V, Vasnetsov M V and Soskin M S 1990 Laser beams with screw dislocations in their wavefronts *JETP Lett.* **52** 429
[28] Leach J, Padgett M J, Barnett S M, Franke-Arnold S and Courtial J 2002 Measuring the orbital angular momentum of a single photon *Phys. Rev. Lett.* **88** 257901
[29] Leach J, Courtial J, Skeldon K, Barnett S M, Franke-Arnold S and Padgett M J 2004 Interferometric methods to measure orbital and spin, or the total angular momentum of a single photon *Phys. Rev. Lett.* **92** 013601
[30] Gao C, Qi X, Liu Y, Xin J and Wang L 2011 Sorting and detecting orbital angular momentum states by using a Dove prism embedded Mach-Zehnder interferometer and amplitude gratings *Opt. Comm.* **284** 48
[31] Karimi E, Piccilillo B, Nagali E, Marrucci L and Santamoto E 2009 Efficient generation and sorting of orbital angular momentum eigenmodes of light by thermally tuned q-plates *Appl. Phys. Lett.* **94** 231124
[32] Slussarenko S, D'Ambrosio V, Piccirillo B, Marrucci L and Santamoto E 2010 The polarizing sagnac interferometer: a tool for light orbital angular momentum sorting and spin–orbit photon processing *Opt. Exp.* **18** 27205

[33] Guo C S, Yue S J and Wei G X 2009 Measuring the orbital angular momentum of optical vortices using a multipinhole plate *Appl. Phys. Lett.* **94** 231104
[34] Padgett M J and Allen L 2002 Orbital angular momentum exchange in cylindrical-lens mode converters *J. Opt. B: Quantum Semiclassical Opt.* **4** S17
[35] Berkhout G C G, Lavery M P J, Courtial J, Beijersbergen M W and Padgett M J 2010 Efficient sorting of orbital angular momentum states of light *Phys. Rev. Lett.* **105** 153601
[36] Lavery M P J, Robertson D J, Berkhout G C G, Love G D, Padgett M J and Courtial J 2012 Refractive elements for the measurement of the orbital angular momentum of a single photon *Opt. Exp.* **20** 2110
[37] Walsh G F 2016 Pancharatnam-Berry optical element sorter of full angular momentum eigenstate *Opt. Exp.* **24** 6689
[38] He H, Heckenberg N R and Rubinsztein-Dunlop H 1995 Optical particle trapping with higher-order doughnut beams produced using high efficiency computer generated holograms *J. Mod. Opt.* **42** 217
[39] He H, Friese M, Heckenberg N R and Rubinsztein-Dunlop H 1995 Direct observation of transfer of angular momentum to absorptive particles from a laser beam with a phase singularity *Phys. Rev. Lett.* **75** 826
[40] Ashkin A, Dziedzic J M and Yamane T 1987 Optical trapping and manipulation of single cells using infrared laser beams *Nature* **330** 769
[41] Sato S, Ishigure M and Inaba H 1991 Optical trapping and rotational manipulation of microscopic particles and biological cells using higher-order mode Nd:YAG laser beams *Electron. Lett.* **27** 1831
[42] Ladavac K and Grier D G 2004 Microoptomechanical pumps assembled and driven by holographic optical vortex arrays *Opt. Exp.* **12** 1144
[43] Vasnetsov M V, Torres J P, Petrov D V and Torner L 2003 Observation of the orbital angular momentum spectrum of a light beam *Opt. Lett.* **28** 2285
[44] Lavery M P J, Speirits F C, Barnett S M and Padgett M J 2013 Detection of a spinning object using lights orbital angular momentum *Science* **341** 537
[45] Rosales-Guzman C, Hermosa N, Belmonte A and Torres J P 2013 Experimental detection of transverse particle movement with structured light *Sci. Rep.* **3** 2815
[46] Lavery M P J, Barnett S M, Speirits F C and Padgett M J 2014 Observation of the rotational Doppler shift of a white-light orbital-angular-momentum-carrying beam backscattered from a rotating body *Optica* **1** 1
[47] Padgett M 2014 A new twist on the Doppler shift *Phys. Today* **67** 58
[48] Jha A K, Agarwal G S and Boyd R W 2011 Supersensitive measurement of angular displacements using entangled photons *Phys. Rev. A* **83** 053829
[49] Führhapter S, Jesacher A, Bernet S and Ritsch-Marte M 2005 Spiral phase contrast imaging in microscopy *Opt. Exp.* **13** 689
[50] Führhapter S, Jesacher A, Maurer C, Bernet S and Ritsch-Marte M 2007 Spiral phase microscopy *Adv. Imag. Electron Phys.* **146** 1
[51] Maurer C, Jesacher A, Führhapter S, Bernet S and Ritsch-Marte M 2008 Upgrading a microscope with a spiral phase plate *J. Microsc.* **230** 134
[52] Larkin K G, Bone D J and Oldfield M A 2001 Natural demodulation of two-dimensional fringe patterns. I. General background of the spiral phase quadrature transform *J. Opt. Soc. Am. A* **18** 1862

[53] Jack B, Leach J, Franke-Arnold S, Ritsche-Marte M, Barnett S M and Padgett M J 2009 Holographic Ghost imaging and the violation of a Bell inequality *Phys. Rev. Lett.* **103** 083602
[54] Torner L, Torres J P and Carrasco S 2005 Digital spiral imaging *Opt. Exp.* **13** 873
[55] Molina-Terriza G, Rebane L, Torres J P, Torner L and Carrasco S 2007 Probing canonical geometrical objects by digital spiral imaging *J. Eur. Opt. Soc.* **2** 07014
[56] Simon D S and Sergienko A V 2012 Two-photon spiral imaging with correlated orbital angular momentum states *Phys. Rev.* A **85** 043825
[57] Uribe-Patarroyo N, Fraine A M, Simon D S, Minaeva O M and Sergienko A V 2013 Object Identification using correlated orbital angular momentum states *Phys. Rev. Lett.* **110** 043601
[58] Fitzpatrick C A, Simon D S and Sergienko A V 2015 High-capacity imaging and rotationally insensitive object identification with correlated orbital angular momentum states *Int. J. Quant. Inf.* **12** 1560013

A Guided Tour of Light Beams
From lasers to optical knots
David S Simon

Chapter 5

Bessel beams, self-healing, and diffraction-free propagation

5.1 Bessel beams

Bessel beams are another beam-like solution to the Helmholtz equation. They have a number of unusual properties; two of the most interesting are that they have a degree of immunity from diffraction and that they can heal themselves after being disrupted by an obstacle. As a result of these features, Bessel beams have been of enormous interest to researchers in recent years.

Bessel beam modes are labelled by an integer n, called the *order* of the beam. In cylindrical coordinates, the electric field for the nth-order Bessel beam is of the form

$$E_n(r, \phi, z) = \mathcal{E}\, e^{ik_z z} J_n(k_r r) e^{\pm in\phi}, \qquad (5.1)$$

where $\mathcal{E}$ is the amplitude. The function $J_n(x)$ is the nth-order Bessel function of the first kind. The definition of this function and its basic properties are given in appendix A.3. Using these properties, the reader can verify by substitution that the expression given above really is a solution to the Helmholtz equation. An additional Gaussian factor can be introduced that decreases the amplitude away from the origin; the beam is then called a *Bessel–Gauss beam*.

The zero-order beam E_0 has an intensity maximum on the axis, like the Gaussian beam; however, unlike the Gaussian beam it also has a set of circular nodes ringing the axis (figure 5.1). But for $n > 0$, a minimum also appears on the central axis. The on-axis node arises for the same reason it did in the Laguerre–Gauss case: the $e^{\pm in\phi}$ factor means that the beam has orbital angular momentum (OAM) $L_z = n\hbar$ about the beam axis, and that the phase of the beam increases as you circle the axis. As a result, the phase is undefined or singular on the axis itself, and therefore the intensity must vanish there.

A true Bessel beam is impossible to create in the lab, since it would need to have infinite extent in the transverse direction, and it would carry infinite power. In this respect, it is similar to a true plane wave, which also has infinite extent and therefore cannot ever be realized exactly. But just as for the plane wave, it is possible to

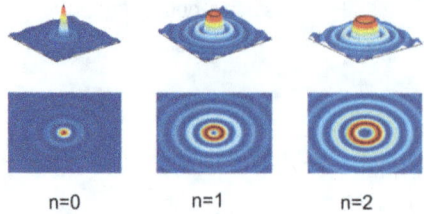

n=0   n=1   n=2

Figure 5.1. Intensity profiles of several Bessel beams in a transverse plane. The figure on the left is for the $n=0$ case, where the axis has an intensity maximum and is ringed by a sequence of circular nodes. In the $n \neq 0$ cases (middle and right) the bright central spot is replaced by an on-axis node. Without the bright spot at the center, the surrounding rings become more intense, so that the total power remains the same. The top row shows each case as viewed from an oblique angle from the axis; the bottom row shows the same plots as viewed from directly along the axis.

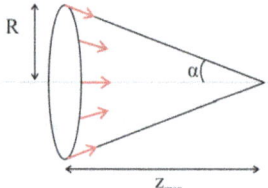

Figure 5.2. The wavevectors for a Bessel beam lie on a cone of opening angle $\alpha = \tan^{-1} \frac{R}{z_{max}}$. R is the aperture angle of the lens and z_{max} is the maximum propagation distance. The wavefronts are perpendicular to the wave vectors and their interference gives rise to the Bessel beam.

produce very good *approximate* Bessel beams which have all of the useful properties of the idealized version, at least over some finite region. It can be shown that each of the bright rings carries the same amount of power; therefore, for a beam of finite power, the intensity of each ring must go down as the number of rings grows. The intensity of the bright central core must also drop as the number of rings increases.

Bessel beams had been described at least as far back as the 1940s [1], but little attention was paid to them because of the need for infinite energy; it wasn't until they were rediscovered 45 years later [2, 3] that people began to take them seriously. Once produced and detected experimentally, a number of applications for them were quickly found.

The key to producing Bessel beams is to create a set of plane waves such that their wavevectors k all lie on a cone (figure 5.2). Recall that the wavefronts are perpendicular to their wavevectors, so that wavefronts corresponding to different k vectors will intersect each other and interfere on the intersection region. The observed beam results from this interference. Here we discuss the beams in the context of optics, but similar conical wave sets have appeared in other areas as well, for example in seismic imaging [4]. We will let α denote the opening angle of the cone, so all wavevectors in the beam are at angle α from the z-axis. The radial and longitudinal components of the wavevectors then obey

$$\tan \alpha = \frac{|k_r|}{k_z}. \qquad (5.2)$$

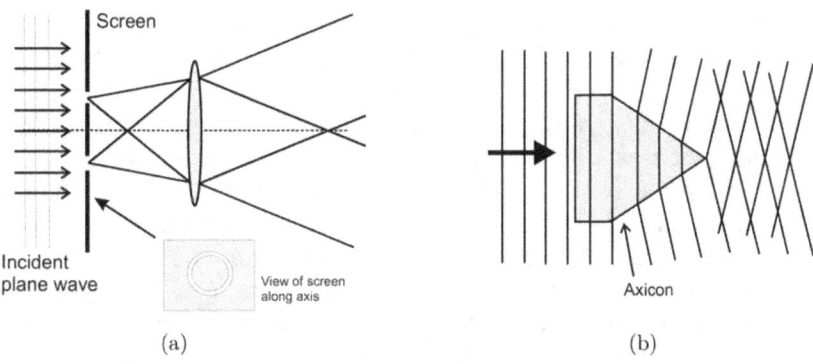

Figure 5.3. (a) A plane wave is sent through an annular aperture and a converging lens it order to create a zero-order Bessel beam. (b) Gaussian beams or Laguerre–Gauss beams can produce zeroth-order or higher-order Bessel beams by passing them through a conical lens called an axicon, which bends all the ougoing wavefronts to a fixed angle.

It is well known that the diffraction pattern from apertures with circular symmetry have amplitudes proportional to Bessel functions (see [5] or any other introductory optics textbook). One simple way to create a Bessel beam is as shown in figure 5.3(a). An annular (ring-shaped) slit in a screen is illuminated with a plane wave P. This annular aperture allows a ring of light to pass, which is then bent inward, toward the axis, by a converging lens of radius R. The lens is placed one focal length f from the aperture. The interference pattern that is formed is then the desired zero-order Bessel beam. Higher order Bessel beams can then be constructed by interferometric means [6].

The use of annular apertures is not very efficient, since most of the light is blocked by the opaque screen that the aperture is cut into. A more efficient method is to create Bessel beams by using an *axicon*, a type of cone-shaped lens (figure 5.3(b)). The inclined sides of the axicon cause an incoming wave-front to be bent inward toward the axis at a fixed angle. If the axicon is illuminated by a Gaussian beam, then the result is a zeroth-order Bessel beam. Illumination of the axicon with Laguerre–Gauss beams can create higher-order Bessel beams, carrying non-trivial angular momentum.

Recall that all light beams tend to spread due to diffraction as they propagate, causing the beam radius to increase. One of the things that makes the Bessel beam of great interest is that it is to a large extent resistant to this diffractive spreading. The beam retains its initial size and shape. For a true Bessel beam of infinite transverse extent, this would be true forever. For realistic approximate Bessel beams, it remains true only up to some maximum distance z_{max}, at which point the beam begins to break up. The initial radius R of the beam is limited by the size of the lens used to create it. Recall that as the radius grows, diffractive effects diminish. So it can be seen from figure 5.2 that the maximum propagation distance is given by

$$z_{max} = \frac{R}{\tan \alpha}. \tag{5.3}$$

If the opening angle α of the propagation cone is made very small, then z_{max} can be large. In the idealized case of a true Bessel beam, $R \to \infty$, the diffraction-free distance becomes infinite, $z_{max} \to \infty$.

Up until the vicinity of z_{max} is reached, the field is approximately translation-invariant along the z-axis. This means that any transverse plane should look like any other transverse plane, regardless of where the planes are located along the axis. This translation-invariance is another way of saying that the beam is diffraction-free; the beam does not spread or become distorted in the transverse direction. This is ultimately due to the fact that all the rays are at the same angle. All rays crossing the axis at a given point have traveled the same distance and picked up the same phase shift. Consequently, there can be no destructive interference and so no alteration of the intensity pattern.

A second feature of Bessel beams that has attracted great interest is that they can reform themselves after being disrupted by an obstacle in the beam path. The mechanism underlying this self-healing property is simple and can be seen in figure 5.4. The obstacle blocks some rays, so that the beam is disrupted in the region just to its right. But after propagating a little farther, rays that missed the obstacle begin to reach the axis and interfere. After a short distance the beam looks exactly the same as before the obstacle. This mechanism relies on the fact that all of the rays (all of the k vectors) are at the same angle from the axis, so that any region that is filled by these rays should look like any other such region, regardless of what has happened before that point. This self-healing occurs in other types of beams as well; an experimental image of self-healing in an Airy beam is shown in figure 5.5. This ability of some beams to reconstruct themselves after an obstacle has a number of applications in microscopy [8] and other areas.

Beams with Bessel-like profile have also been found [9] that bend (or self-accelerate) as they propagate. The concept of self-acceleration will be discussed in chapter 6.

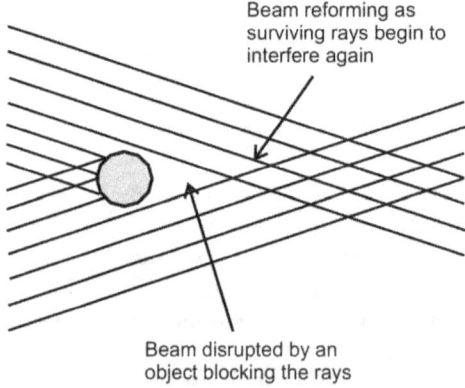

Figure 5.4. Self-healing of the Bessel beam after an obstacle. The beam is traveling toward the right.

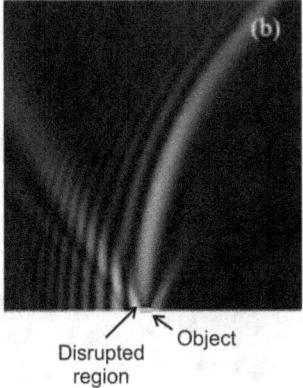

Figure 5.5. An experimental image showing the self-healing phenomenon. The beam is moving upward in the picture, and at the bottom it has encountered an opaque obstacle (the small green rectangle). The beam is disrupted just after the obstacle, but quickly reforms and returns to its original shape as rays that leaked around the obstacle begin to reach the vicinity of the axis. (The beam shown here is actually a non-diffracting Airy beam rather than a Bessel beam. Figure reproduced from [7].)

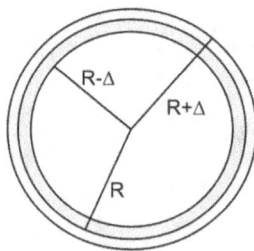

Figure 5.6. A split-ring aperture, formed by two ring-shaped apertures in an opaque screen. Each is of thickness Δ and the boundary between them has radius R, with $R \gg \Delta$. Each ring has a piece of glass or other material of azimuthally-varying thickness, in order to give non-zero orbital angular momentum to the light passing through each ring. The result is a superposition of two higher-order Bessel beams with different OAM.

5.2 Optical petal structures

Now consider [10] an opaque screen with two thin annular apertures, one just outside the other, and each of width Δ, as in figure 5.6. We can approximate them as both having the same radius, R, as long as $\Delta \ll R$. One aperture is arranged to add an azimuthal phase shift $e^{im\phi}$, while the other adds a shift $e^{in\phi}$. This is called a *split-ring aperture*. Each aperture separately produces a higher-order Bessel beam, one of orbital angular momentum m, the other of OAM n. The two together will then produce a superposition of these beams:

$$E_{m,n}(r, \phi, z) = E_m(r, \phi, z) + E_n(r, \phi, z). \tag{5.4}$$

The resulting flower-like interference pattern has rings, like a Bessel function; but instead of being continuous, each ring consists of $|n - m|$ bright and dark spots forming a petal structure, as in figure 5.7. Using some trigonometric identities and

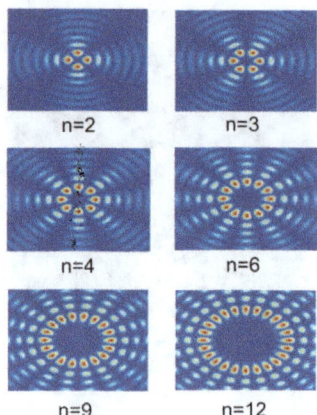

Figure 5.7. Intensity profiles in the transverse plane for superpositions of Bessel beams formed by split-ring apertures. For the cases plotted here, the two terms in the superposition have equal and opposite angular momenta, $m = -n$, so that the total angular momentum in each case vanishes.

the Euler formula, the phase factors of the two fields in the superposition combine as follows:

$$e^{im\phi} + e^{in\phi} = (\cos m\phi + \cos n\phi) + i(\sin m\phi + \sin n\phi) \tag{5.5}$$

$$= 2\cos\frac{(m+n)\phi}{2}\cos\frac{(m-n)\phi}{2} + 2i\sin\frac{(m+n)\phi}{2}\cos\frac{(m-n)\phi}{2}$$
$$= 2\cos\frac{(m-n)\phi}{2}\,e^{i\frac{(m+n)\phi}{2}}. \tag{5.6}$$

So the difference in the *phases* of the two terms causes the *amplitude* to vary sinusoidally around the circle due to the cosine term, accounting for the bright and dark spots on each ring. The number of bright and dark spots depends on the difference, $|m - n|$. In addition, the exponential factor tells us that the *phase* is varying linearly around each ring, or in other words that the combined field has OAM $L_z = \frac{1}{2}(m+n)\hbar$. As the field propagates along the axis, the entire pattern rotates, since there is also a phase e^{-ikz} proportional to the distance travelled (see equation (5.1)).

In the case where $m = -n$, the pattern still rotates, even though the superposition has a net orbital angular momentum of zero. This is because the light leaves the two halves of the aperture with slightly different z-components of momentum, due to the slightly different angles subtended by the two annuli. Because their longitudinal momenta are slightly mismatched, there is an overall phase difference between the two Bessel beams, and this phase difference changes at a constant rate as z increases, leading to a rotation of the entire interference pattern.

5.3 More non-diffracting beams: Mathieu beams

Bessel beams are not the only type of wave solution that resists diffraction. There is in fact a whole family of diffraction-free beams [11], all of which are formed from plane waves whose wavevector components lie on a circle, $\boldsymbol{k} = k_\perp(\cos\phi\,\hat{x} + \sin\phi\,\hat{y}) + k_z\hat{z}$. Each transverse component of the electric field can be written in the form of Whittaker's integral [12, 13],

$$E(\boldsymbol{r}) = \int_0^{2\pi} d\phi\, f(\phi) e^{-i\boldsymbol{k}\cdot\boldsymbol{r}} = e^{-ik_z z} \int_0^{2\pi} d\phi\, f(\phi) e^{-ik_\perp(x\cos\phi + y\sin\phi)}, \tag{5.7}$$

for some function $f(\phi)$ of the azimuthal angle. Different subgroups of non-diffracting beams inside this general class are distinguished by having different angular distribution functions, $f(\phi)$. Bessel beams, for example, fall into this category with distribution function $f(\phi) = e^{il\phi} = \cos(l\phi) + i\sin(l\phi)$. The possible Whittaker-type classes of non-diffracting beams are summarized in table 5.1. Each solution in this class is separable in a particular coordinate system. Separation of variables in various coordinate systems is discussed in chapter 7.

A second class of Whittaker-type non-diffracting beams are the *Mathieu beams*, characterized by the distribution function $f(\phi) = C(l, q, \phi) + iS(l, q, \phi)$, where the angular Mathieu sine and cosine functions, $C(l, q, \phi)$ and $S(l, q, \phi)$, are elliptical generalizations of the usual circular sine and cosine functions. (The functions $C(l, q, \phi)$ and $S(l, q, \phi)$ are also sometimes denoted $ce_l(q, \phi)$ and $se_l(q, \phi)$.) The parameter q is the ellipticity parameter, and as $q \to 0$ these reduce to the usual trigonometric functions:

$$C(l, 0, \phi) = \cos l\phi, \qquad S(l, 0, \phi) = \frac{\sin l\phi}{l}. \tag{5.8}$$

C and S are even and odd solutions, respectively, to the Mathieu differential equation,

$$\frac{d^2 y}{d\phi^2} + \left(l^2 - 2q\cos(2\phi)\right) y = 0, \tag{5.9}$$

Table 5.1. Table of the possible Whittaker-type non-diffracting beams. Each is separable in a particular coordinate system and can be expressed in the form of equation (5.7) with the given angular spectrum $f(\phi)$. Parabolic beams are discussed in chapter 7.

Type of solution	Coordinate system	Angular spectrum
Plane wave at angle ϕ_0	Cartesian	$f_p(\phi) = \delta(\phi - \phi_0)$
Bessel	Circular cylindrical	$f_b(\phi) = e^{im\phi}$
Mathieu	Elliptical cylindrical	$f_m(\phi) = C(l, q, \phi) + iS(l, q, \phi)$
Parabolic (even order)	Parabolic cylindrical	$f_e(\phi) = \frac{1}{2(\pi\lvert\sin\phi\rvert)^{1/2}} e^{ia\ln\lvert\tan\phi/2\rvert}$
Parabolic (odd order)	Parabolic cylindrical	$f_o(\phi) = \begin{cases} if_e(\phi), & \text{if } \phi \in (-\pi, 0) \\ -if_e(\phi), & \text{if } \phi \in (0, \pi) \end{cases}$

which can be obtained by writing the Helmholtz equation in terms of elliptical coordinates and then using separation of variables. Elliptical coordinates (ξ, ϕ) are related to Cartesian (x, y) coordinates by

$$x = c \cosh \xi \cos \phi, \qquad y = c \sinh \xi \sin \phi, \tag{5.10}$$

where c is the distance between the foci of the ellipse, ξ is the radial variable, $0 \leq \xi < \infty$, and ϕ is an angular variable, $0 \leq \phi < 2\pi$.

Properties and explicit expressions for the Mathieu functions may be found, for example, in [1, 13–15]. These functions are built into Mathematica under the names MathieuC [a, q, z] and MathieuS[a, q, z]. Toolboxes have also been written to compute them in MATLAB® and can be easily found online. The Mathieu functions can be written as infinite series, summed over the usual sine and cosine functions with different periods,

$$C(2n, q, \phi) = \sum_{r=0}^{\infty} A_{2r}^{(2n)} \cos 2r\phi \tag{5.11}$$

$$C(2n + 1, q, \phi) = \sum_{r=0}^{\infty} A_{2r+1}^{(2n+1)} \cos(2r + 1)\phi \tag{5.12}$$

$$S(2n + 1, q, \phi) = \sum_{r=0}^{\infty} B_{2r+1}^{(2n+1)} \sin(2r + 1)\phi \tag{5.13}$$

$$S(2n + 2, q, \phi) = \sum_{r=0}^{\infty} B_{2r+2}^{(2n+2)} \sin(2r + 2)\phi, \tag{5.14}$$

where the coefficients $A_r^{(n)}$ and $B_r^{(n)}$ have forms that depend on the value of q and obey a set of recurrence relations that can be found listed in [14]. Expanded in a Taylor series about $q = 0$, the first Mathieu sine and cosine function are of the forms [16]:

$$C(1, q, \phi) = \cos \phi - \frac{1}{8} \cos(3\phi)q - \frac{1}{64}\left(-\cos 3\phi + \frac{1}{3} \cos 5\phi\right)q^2$$
$$- \frac{1}{512}\left(\frac{1}{3} \cos 3\phi - \frac{4}{9} \cos 5\phi + \frac{1}{18} \cos 7\phi\right)q^3 + \cdots \tag{5.15}$$

$$S(1, q, \phi) = \sin \phi - \frac{1}{8} \sin(3\phi)q + \frac{1}{64}\left(\sin 3\phi + \frac{1}{3} \sin 5\phi\right)q^2$$
$$- \frac{1}{512}\left(\frac{1}{3} \sin 3\phi + \frac{4}{9} \sin 5\phi + \frac{1}{18} \sin 7\phi\right)q^3 + \cdots \tag{5.16}$$

The Mathieu functions arise in a number of different types of physical problems, ranging from the quantum mechanical pendulum to the vibrational modes of an elliptical membrane; surveys of physical applications may be found in [17, 18].

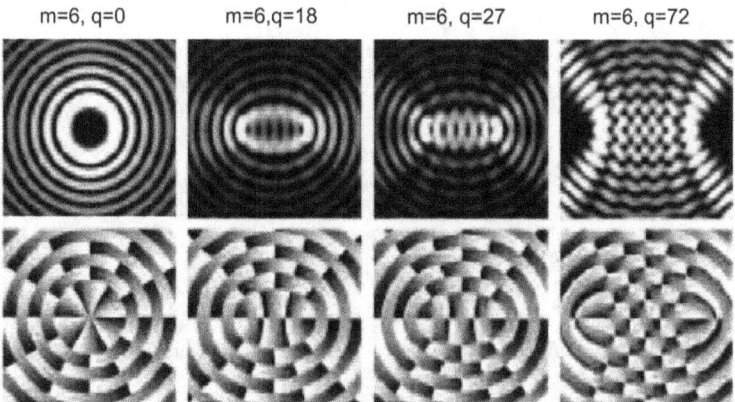

Figure 5.8. Images of the amplitude (top row) and phase (bottom row) of the Mathieu beams for several values of ellipticity q and order m. Notice the multiple singular points in the phase structure, where phase regions of different phase values join; the amplitude must vanish at those points. The presence of multiple phase singularities indicates a complicated OAM structure in the beam. (Figure reproduced from [22].)

Plugging the angular Mathieu functions into the Whittaker integral leads to the non-diffractive Mathieu beam modes. Examples of Mathieu beam cross-sections are shown in figure 5.8. Like Bessel beams, the Mathieu beams also have the self-healing property. Furthermore, there are variations on the Mathieu beam that can seem to bend [19], exhibiting self-acceleration; see chapter 6.

Mathieu beams can be represented as a sum of Bessel beams, and so they can be generated by similar methods [20]. For example, plane waves may be passed through a transparency with a prescribed spatial transmission profile, followed by passage through an annular aperture and a lens. The main difference is in the angular form of the transmission profile, which is given by the angular Mathieu functions [1, 21].

A holographic generation method can also be carried out [22], using computer-generated holograms. This is based on the fact that Mathieu beams can be represented as superpositions of Bessel beams, which in turn have been successfully generated holographically [23].

References [7, 24] contain detailed reviews of a number of types of non-diffracting beams and waves. It should also be pointed out that in addition to directed beams, there has been related work on diffraction-free propagation of extended optical patterns and images [25].

5.4 Optical tractor beams and conveyor belts

Normally, if a beam of light is travelling to the right, then any object in the beam path will also be pushed to the right by radiation pressure. The photons making up the beam exert pressure in the direction of their momentum. On Star Trek you may have seen *tractor beams*, beams of radiation that pull objects *toward* the radiation source (the Enterprise) instead of pushing them *away*. Until recently, this sort of phenomenon has only existed in science fiction, but in the past few years it has been implemented on a microscopic scale using Bessel beams. There are several possible

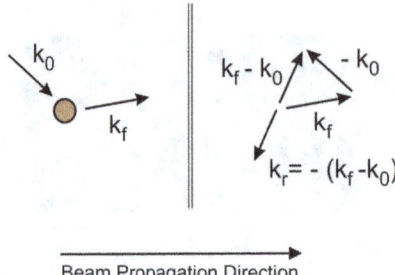

Figure 5.9. One mechanism for an optical tractor beam. A Bessel beam photon with initial momentum k_0 strikes a particle, and scatters off it with final momentum k_f, as in the figure on the left. On the right, the momentum vectors are drawn: in order to conserve momentum, the particle must gain a recoil momentum $k_r = -(k_f - k_0)$ opposite to the momentum change of the photon. This recoil momentum has a component pointing backwards, toward the source of the light. The transverse part of the recoil momentum will cancel on average, since every k_0 on the cone of figure 5.2 has a partner on the other side of the cone with opposite transverse component.

mechanisms for doing this. The simplest uses the fact that if the light has a momentum component in the transverse direction, then when the light scatters off the object, momentum conservation forces the object to gain a recoil momentum that can have a component opposite to the overall direction of the beam (figure 5.9). A related mechanism allows the construction of *optical conveyor belts* that can use optical forces to drag a particle along some prescribed path by forcing it to hop from one optical trap to another [26]. Such a mechanism is potentially very useful for constructing nanostructures by using beams of light to manipulate the building blocks, for engineering biological tissues to repair damaged natural tissue, and for other uses. For more detail, see [27–30] and references therein. Some of the work in these references actually involves acoustic waves, rather than light, but the mechanism is the same regardless of the wave type.

5.5 Trojan states

Another potential application of Bessel beams is to guide individual electrons along a desired path by means of optical forces. The electrons are smuggled through space inside the light beam like Trojan soldiers inside Homer's wooden horse.

The Trojan asteroids orbit the Sun in the same orbit as Jupiter, at a pair of points (called the *Lagrange points* or *L*4 and *L*5 points), where the gravitational effects of the Sun and Jupiter conspire to create stable equilibria. Replacing the gravitational forces by electromagnetic forces, a microscopic analog of this effect exists. Consider an atomic electron sitting in a highly excited state, with high principal quantum number; such a state is called a *Rydberg state*. The resulting Rydberg atom is then subjected to polarized microwaves, where it is arranged so that the photon polarization rotates in phase with the electronic orbital period. The electron can then be made to orbit the atom so that it remains tightly-localized, without spreading [31]. This is in contradiction to the conclusion reached in the early days of quantum mechanics by Schrödinger and Lorentz [32], who believed that the electron wavefunction should become uniformly spread over the orbit. Such stable, non-spreading electronic states are called *Trojan states*, in analogy with the behavior of the Trojan asteroids.

These atomic Trojan states generally follow circular orbits. But by inserting electrons into optical beams it is also possible, at least in principle, to make them follow optical vortex lines of the beam without spreading. It has been shown theoretically [33, 34] that higher order Bessel beams can be used as guides to direct electrons to a desired location or along a desired path. The Lorentz and Coriolis forces conspire to keep the electrons spiraling and wiggling around the beam axis without straying too far from the central vortex. The mechanism which keeps the electrons confined to the vicinity of the beam is similar to that of the circular atomic Trojan states. Numerical estimates [34] show that beams of moderate intensity ($\sim 10^{14}$ W m^{-2}) are capable of confining electrons with speeds of over 10^5 m s^{-1}.

5.6 Localized waves

In the previous section, atomic Trojan states were mentioned, where the electrons are prevented from spreading by a containing light beam. It is also possible for the beam to effectively exert a containing influence on itself.

Since the 1980s a number of other diffraction-free solutions to Maxwell's equations have been derived, besides those of Whittaker-type. In particular, a number of solutions have been found that are highly localized in all three dimensions, and which do not spread appreciably as they propagate through free space. In other words, the wave acts as a localized particle or bullet. Like the Bessel beam, these waves are ideally diffraction-free. The first such solution found [35] was quickly followed by a stream of work [36–47] that led to the discovery of other localized solutions. The catch is that in order to be truly non-diffracting, the aperture used to produce them has to be infinite in size; the solutions that can be practically realized with finite-size equipment therefore do spread slowly. But they can be arranged to travel for very long distances before any appreciable change in their profile, acting as bullets or pulses of light that remain roughly the same size as they propagate.

Wave equation solutions that have similar properties have long been known. These solutions are called *solitons*. Solitons are ultimately of topological origin, and have been thoroughly studied in areas such as fluid dynamics and elementary particle physics [48–51]. The soliton was discovered by a Scottish engineer, John Scott Russell, in 1834. Russell spotted an odd-looking wave moving along the surface of a canal. The wave had been created when a boat stopped suddenly, and it seemed to propagate without noticeably spreading or dissipating. He followed it on horseback as it moved down the canal, until he lost sight of it more than a mile later. Before losing it, he saw that in fact the height of the wave was diminishing, but only at a very slow rate (so it was only an *approximate* soliton). Solitons have been well-studied in the mathematical literature since the late 19th century, and are now well-understood. It is well-established that *nonlinear* wave equations are required to produce a soliton solution.

However, the localized optical solutions mentioned above occurred in free space, where the wave equation is completely linear; therefore, they could not be solitons in the strict sense and were in fact a new type of wave solution. (Optical solitons do exist, but require the presence of nonlinear optical materials; see [52] for a review.)

One type of such localized, limited-diffraction wave is the *X-wave* or *bowtie wave* [53, 54], so called because of the shape of their amplitude profile, figure 5.10. They can be defined in terms of Laplace transforms or as non-monochromatic superpositions of Bessel beams, and can all be written as derivatives of a single generating function. They can be generated by means of axicons or through nonlinear processes, and approximate X-waves have been experimentally observed [55]. Quantum states with X-wave profile can also be constructed [56].

One interesting feature of these X-wave states is that multiple experiments [57, 58] have shown them to have group velocities exceeding the speed of light. There has been much discussion about these experiments and about how they may be reconciled with special relativity [59–61]; although it is agreed that information-bearing signals must travel at less than the speed of light, there still seems to be some

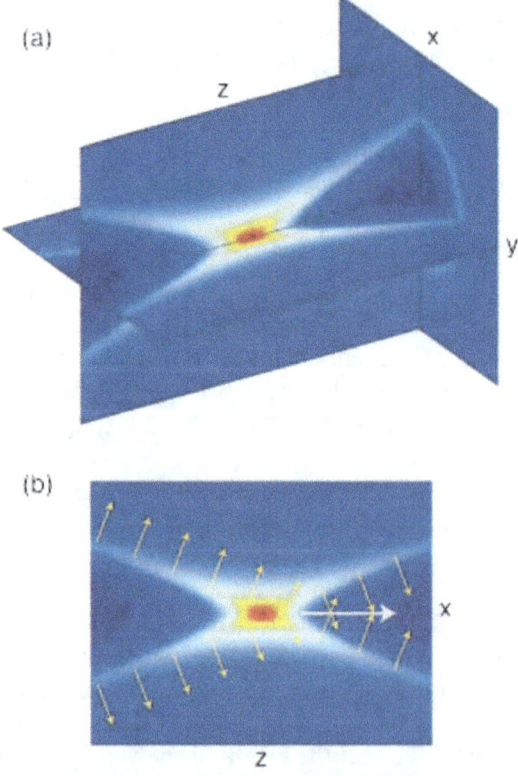

Figure 5.10. An example of an X-wave. The small yellow arrows display the propagation direction of the energy flow and the wavefronts, the large arrow gives the propagation direction of the interference pattern as a whole. (Figure reproduced from [65].)

debate about what actually constitutes a 'signal' [62–64]. It should be noted that the individual photons in the beam travel at the speed of light; it is the *interference fringes* formed by their wavefronts that move with superluminal speeds.

A unified description of X-waves and the other localized known wave solutions can be constructed using Fourier transforms [65].

Bibliography

[1] Stratton J A 1941 *Electromagnetic Theory* (New York: McGraw-Hill)
[2] Durnin J 1987 Exact solutions for nondiffracting beams. I. The scalar theory *J. Opt. Soc. Am.* A **4** 651
[3] Durnin J, Miceli J J and Eberly J H 1987 Diffraction-free beams *Phys. Rev. Lett.* **58** 1499
[4] Nowack R L 2012 A tale of two beams: an elementary overview of Gaussian beams and Bessel beams *Stud. Geophys. Geod.* **56** 355
[5] Hecht E 2016 *Optics* 5th edn (Upper Saddle River, NJ: Pearson)
[6] López-Mariscal C, Gutiérrez-Vega J C and Chávez-Cerda S 2004 Production of high-order Bessel beams with a Mach-Zehnder interferometer *Appl. Opt.* **43** 5060
[7] Mazilu M, Stevenson D J, Gunn-Moore F and Dholakia K 2010 Light beats the spread: 'non-diffracting' beams *Laser Phot. Rev.* **4** 529
[8] Fahrbach F O, Simon P and Rohrbach A 2010 Microscopy with self-reconstructing beams *Nat. Phot.* **4** 780
[9] Chremmos I D and Efremidis N K 2013 Nonparaxial accelerating Bessel-like beams *Phys. Rev.* A **88** 063816
[10] Vasilyeu R, Dudley A, Khilo N and Forbes A 2009 *Opt. Exp.* **17** 23389
[11] Salo J, Fagerholm J, Friberg A T and Salomaa M M 2000 Unified description of nondiffracting X and Y waves *Phys. Rev.* E **62** 4261
[12] Whittaker E T 1903 On the partial differential equations of mathematical physics *Math. Ann.* **57** 333
[13] Whittaker E T and Watson G 1927 *A Course of Modern Analysis* (Cambridge: Cambridge University Press)
[14] Gradshteyn I S and Ryzhik I M 2007 *Table of Integrals, Series, and Products* ed A Jeffrey and D Zwillinger 7th edn (London: Academic)
[15] Weisstein E E 2003 *CRC Concise Encyclopedia of Mathematics* 2nd edn (Boca Raton, FL: Chapman and Hall/CRC)
[16] Arfken G, Weber H and Harris F E 2012 *Mathematical Methods for Physicists: a Comprehensive Guide* 7th edn (London: Academic)
[17] Ruby L 1996 Applications of the Mathieu equation *Am. J. Phys.* **64** 39
[18] McLachlan N W 1947 *Theory and Applications of Mathieu Functions* (Oxford: Clarendon)
[19] Zhang P, Hu Y, Li T, Cannan D, Yin X, Morandotti R C and Zhang X 2012 Nonparaxial mathieu and weber accelerating beams *Phys. Rev. Lett.* **109** 193901
[20] Guttiérez-Vega J C, Iturbe-Castillo M D, Ramírez S G A, Tepichín E, Rodriguez-Dagnino R M and Chávez-Cerda S 2001 Experimental demonstration of optical mathieu beams *Opt. Comm.* **195** 35
[21] Guttiérez-Vega J C, Iturbe-Castillo M D and Chávez-Cerda S 2000 Alternative formulation for invariant optical fields: Mathieu beams *Opt. Lett.* **25** 1493

[22] Chávez-Cerda S, Padgett M J, Allison I, New G H C, Guttiérez-Vega J C, O'Neil A T, MacVicar I and Courtial J 2002 Holographic generation and orbital angular momentum of high-order Mathieu beams *J. Opt.* B **4** S52
[23] Turunen J, Vasara A and Friberg A T 1988 Holographic generation of diffraction-free beams *Appl. Opt.* **27** 3959
[24] Hernández-Figueroa H E, Recami E and Zamboni-Rached M (ed) 2014 *Non-Diffracting Waves* (Weinheim: Wiley-VCH)
[25] Bouchal Z 2002 Controlled spatial shaping of nondiffracting patterns and arrays *Opt. Lett.* **27** 1376
[26] Hansen P, Zheng Y, Ryan J and Hesselink L 2014 Nano-optical conveyor belt, part I: theory *Nano Lett.* **14** 2965
[27] Marston P L 2006 Axial radiation force of a Bessel beam on a sphere and direction reversal of the force *J. Acoust. Soc. Am.* **120** 3518
[28] Chen J, Ng J, Lin Z and Chan C T C T 2011 Optical pulling force *Nature Photon* **5** 531
[29] Ruffner D B and Grier D G 2012 *Phys. Rev. Lett.* **109** 163903
[30] Mitri F G 2014 Single Bessel tractor-beam tweezers *Wave Motion* **51** 986
[31] Maeda H, Gurian J H and Gallagher T F 2009 Nondispersing bohr wave packets *Phys. Rev. Lett.* **102** 103001
[32] Przibram K (ed) 1967 *Letters on Wave Mechanics* (New York: Philosophical Library)
[33] Bialynicki-Birula I 2004 Particle beams guided by electromagnetic vortices: new solutions of the Lorentz, Schrödinger, Klein–Gordon, and Dirac equations *Phys. Rev. Lett.* **93** 20402
[34] Bialynicki-Birula I, Bialynicki-Birula Z and Chimura B 2005 Trojan states of electrons guided by bessel beams *Laser Phys.* **15** 1371
[35] Brittingham J B 1983 Focus wave modes in homogeneous Maxwell's equations: transverse electric mode *J. Appl. Phys.* **54** 1179
[36] Ziolkowski R W 1985 Exact solutions of the wave equation with complex source locations *J. Math. Phys.* **26** 861
[37] Ziolkowski R W, Lewis D K and Cook B D 1989 Evidence of localized wave transmission *Phys. Rev. Lett.* **62** 147
[38] Shaarawi A M, Besieris I M and Ziolkowski R W 1989 Localized energy pulse train launched from an open, semi-infinite, circular waveguide *J. Appl. Phys.* **65** 805
[39] Besieris I M, Shaarawi A M and Ziolkowski R W 1989 A bidirectional traveling plane wave representation of exact solutions *J. Math. Phys.* **30** 1254
[40] Heyman E, Steinberg B and Felsen L B 1987 Spectral analysis of focus wave modes *J. Opt. Soc. Am.* A **4** 2081
[41] Ziolkowski R W 1989 Localized transmission of electromagnetic energy *Phys. Rev.* A **39** 2005
[42] Candy J V, Ziolkowski R W and Lewis D K 1990 Transient wave estimation: a multichannel deconvolution application *J. Acoust. Soc. Am.* **88** 2235
[43] Hillion P 1990 The Goursat problem for the homogeneous wave equation *J. Math. Phys.* **31** 1939
[44] Hillion P 1990 The Goursat problem for Maxwells equations *J. Math. Phys.* **31** 3085
[45] Sezginer A 1985 A general formulation of focus wave modes *J. Appl. Phys.* **57** 678
[46] Palmer M and Donnelly R 1993 Focused waves and the scalar wave equation *J. Math. Phys.* **34** 4007

[47] Borisov B B and Utkin A B 1994 Some solutions of the wave and Maxwell's equations *J. Math. Phys.* **35** 3624
[48] Rajaraman R 1982 *Solitons and Instantons: An Introduction to Solitons and Instantons in Quantum Field Theory* (Amsterdam: Elsevier)
[49] Drazin P G and Johnson R S 1989 *Solitons: An Introduction* (Cambridge: Cambridge University Press)
[50] Remoissenet M 1999 *Waves Called Solitons: Concepts and Experiments* (Berlin: Springer)
[51] Infeld E and Rowlands G 2000 *Nonlinear Waves, Solitons and Chaos* 2nd edn (Cambridge: Cambridge University Press)
[52] Hasegawa A and Kodama Y 1995 *Solitons in Optical Communications* (Oxford: Oxford University Press)
[53] Lu J-Y and Greenleaf J F 1992 Nondiffracting X-waves: Exact solutions to free-space scalar wave equation and their finite aperture realizations *IEEE Trans. Ultrason. Ferroel. Frequation Control* **39** 19
[54] Lu J-Y and Greenleaf J F 1992 Experimental verification of nondiffracting X-waves *IEEE Trans. Ultrason. Ferroel. Frequation Control.* **39** 441
[55] Saari P and Reivelt K 1997 *Phys. Rev. Lett.* **79** 4135
[56] Ciattoni A and Conti C 2007 *J. Opt. Soc. Am.* B **24** 2195
[57] Mugnai D, Ranfagni A and Ruggeri R 2000 Observation of Superluminal Behaviors in Wave Propagation *Phys. Rev. Lett.* **84** 4830
[58] Valtna-Lukner H, Bowlan P, Löhmus M, Piksarv P, Trebino R and Saari P 2009 *Opt. Exp.* **17** 14948
[59] Ringermacher H and Mead L 2001 Comment on observation of superluminal behaviors in wave propagation *Phys. Rev. Lett.* **87** 059402
[60] Bigelow N P and Hagen C R 2001 Comment on observation of superluminal behaviors in wave propagation *Phys. Rev. Lett.* **87** 059401
[61] Recami E, Zamboni-Rached M and Hernandez-Figueroa H E 2008 *Localized Waves* ed H E Hernandez-Figueroa, M Z Rached and E Recami (Hoboken, NJ: Wiley)
[62] Chiao R Y and Steinberg A M 1997 Tunneling times and superluminality *Progress in Optics* Vol 37 (New York: Elsevier)
[63] Brillouin L 1960 *Wave Propagation and Group Velocity* (New York: Academic)
[64] Nimtz G and Heitmann W 1997 Superluminal photonic tunneling and quantum electronics *Prog. Quantum Electron.* **21** 81
[65] Salo J, Fagerholm J, Friberg A T and Salomaa M M 2000 Unified description of nondiffracting X and Y waves *Phys. Rev. E* **62** 4261

IOP Concise Physics

A Guided Tour of Light Beams
From lasers to optical knots
David S Simon

Chapter 6

Airy beams and self-acceleration

6.1 Airy beams

Light beams normally propagate along a straight line, expanding as they travel. But in fact it is possible to make light beams that bend as they propagate, tracing out parabolic, elliptical, or circular trajectories. These beams also exhibit self-healing and the absence of diffraction. The best-known of these so-called self-accelerating beams is the Airy beam, which was first described by Berry and Balazs in 1979. Originally, Airy solutions were discovered [1] as wavefunctions forming a class of solutions to the Schrödinger equation; however, it was soon realized that, due to the similarity in form between the Schrödinger and Helmholtz equations discussed in section 2.2, solutions of the same form appear in optics. The optical version of the Airy beam was first produced experimentally in 2007 [2].

Write the electric field of an optical beam in Cartesian coordinates as

$$E(\mathbf{r}) = u(x, y, z)\, e^{i(\omega t - kz)}, \tag{6.1}$$

and suppose that the envelope $u(x, y, z)$ is given by a slowly-varying function along the z-axis. Then neglecting the z-derivatives of u compared to the z-derivatives of the exponential, the Helmholtz equation leads to the relation:

$$\frac{\partial^2 u}{\mathrm{d}x^2} + \frac{\partial^2 u}{\mathrm{d}y^2} + 2ik\frac{\omega^2}{c^2}u = 0. \tag{6.2}$$

Note that this equation is nonlinear in the wavenumber, since the last term can be written as $2ik^3 u$. This equation can be solved. The most general possible solution is:

$$u(\mathbf{r}) = \int_{-\infty}^{\infty} \mathrm{d}k_y\, \mathrm{Ai}\!\left(\frac{x}{x_0} + x_0^2 k_y^2 - \frac{z^2}{4k^2 x_0^4}\right) \\ \times f(k_y)\exp\!\left(-iky + \frac{ixz}{2kx_0^3} - \frac{iz^3}{12k^3 x_0^6}\right), \tag{6.3}$$

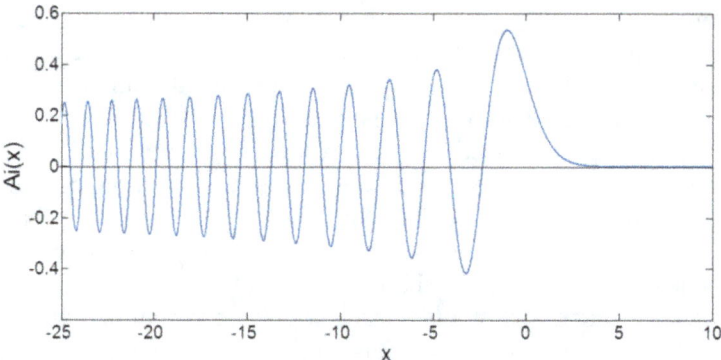

Figure 6.1. The Airy function. This function, which often appears in interference and diffraction problems, is oscillatory for $x < 0$ and decaying for $x > 0$.

for an arbitrary function f. Any rotation of this function in the x–y plane is also a valid solution. Here f is any complex function of the y-component of the momentum, and Ai is the Airy function. For real x, Ai(x) is given by

$$\text{Ai}(x) = \frac{1}{\pi} \int_0^\infty \cos\left(\frac{t^3}{3} + xt\right) dt. \tag{6.4}$$

The Airy function appears in the diffraction pattern of a circular aperture. It is oscillatory for negative x, and exponentially decaying for $x > 0$ (figure 6.1).

The Airy beam is obtained from equation (6.3) by choosing the arbitrary function in the integral to be the Dirac delta function: $f(k_y) = \delta(k_y)$. This reduces the solution to

$$u(r) = \text{Ai}\left(\frac{x}{x_0} - \frac{z^2}{4k^2 x_0^4}\right) \exp\left(\frac{ixz}{2kx_0^3} - \frac{iz^3}{12k^3 x_0^6}\right). \tag{6.5}$$

This describes a beam that seems to bend along a parabolic path in the x–z plane, as discussed in the next section. Other self-accelerating beams based on non-Airy profiles have also been studied theoretically and produced experimentally [3].

Airy beams may be generated by imposing a cubic phase profile on top of a standard Gaussian beam (see equation (6.3)). This can be done, for example, by means of a spatial light modulator (SLM) (see chapter 9).

Although the Airy beam has a total orbital angular momentum of zero, it has been noted [4] that there is a sort of separation of the angular momenta into different regions of the beam in a manner similar to the charge separation of an electrically neutral, but polarized, molecule. The main central peak and the tail both have non-zero angular momentum, which together sum to zero.

6.2 Self-accelerating beams and optical boomerangs

The most unusual aspect of Airy beams is that their paths seem to bend: whereas the axis of a Gaussian beam always follows a perfectly straight line (assuming a perfectly homogeneous propagation medium), Airy beams seem to follow parabolic paths. Such beams are referred to as *self-accelerating*, since their center seems to follow a nonlinear path, accelerating perpendicular to the initial beam direction, despite the absence of external forces acting on the moving photons. Self-acceleration is a property shared by some types of Mathieu and Weber [3] beams as well; in fact, the Airy beam can be viewed as a paraxial special case of the Weber beam. Self-accelerating non-paraxial Mathieu and Weber beams can bend backward so strongly, that they have been referred to as *optical boomerangs*. Non-paraxial self-accelerating beams with Bessel-like shapes also been found [5]. Examples of Mathieu beams following circular and elliptical trajectories are shown in figure 6.2.

For the case of quantum wavefunctions, this acceleration in the absence of forces initially seems to violate *Ehrenfest's theorem* from quantum mechanics. Ehrenfest's theorem states that the expectation values of quantum operators should obey the same equations of motion as the corresponding classical variables. In this case, it means that the mean position (the expectation value) of the beam center, $\langle x(t) \rangle = \langle \psi(t) | \hat{x} | \psi(t) \rangle$, should obey the classical equation of a particle moving freely in the absence of forces:

$$\ddot{x}_{cl} = 0 \iff \frac{d^2 \langle x \rangle}{dt^2} = 0. \tag{6.6}$$

The resolution to this apparent paradox was already pointed out by Berry and Balazs in their original paper [1] on Airy wavefunctions: the wavefunction is not square integrable, or in other words $\int |\psi(r)|^2 d^2x$ diverges. This means that the wavefunction cannot describe the probability density of a single particle, but rather it describes a family of particles, each travelling in a straight line. In the optical case, the field can be thought of as describing an infinite set of rays. When these rays are all drawn, as in figure 6.3, the mechanism for the self-acceleration becomes clear: what is bending is the *caustic* or envelope of the collection of rays. An alternative

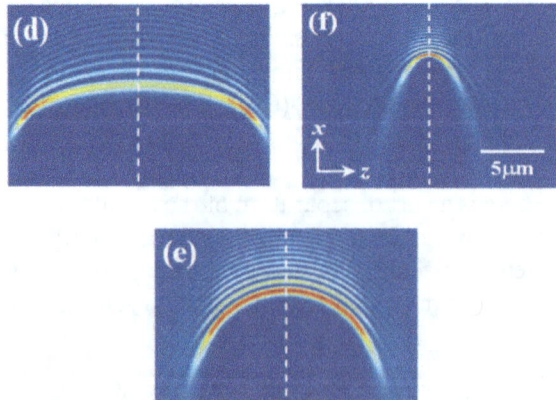

Figure 6.2. Non-paraxial self-accelerating Mathieu beams. The beams can be made to follow elliptical (top row) or circular (bottom) trajectories. (Figures reproduced from [3].)

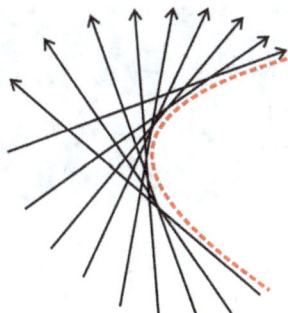

Figure 6.3. The rays of light in an Airy beam cross each other, staying on one side of an enveloping curve (the dashed red curve). Each of the rays is briefly tangent to the envelope at some point. Such an envelope of a system of rays is called an optical caustic.

explanation for the self-accelerating nature of the solutions can be given in terms of the equivalence principle [6].

Caustics are common features in everyday life. For example, place a glass of water on a table, with sunlight illuminating it at a low angle from the horizontal. (The glass has to be clear.) On the tabletop you will see curved, brightly illuminated regions appearing, formed by the envelopes of the refracted light rays. Rainbows are another example of a caustic, produced by the rays refracted by water droplets in the air.

The fact that $\int |\psi(r)|^2 d^2x$ is infinite implies that the energy required to create a true Airy wave is infinite. Therefore, like Bessel beams and plane waves, they are abstractions which can be implemented only approximately in the real world. The lack of square-integrability follows from the so-called 'weak confinement' property of the Airy function, which decays at large distances, but only very slowly, $|\text{Ai}(x)| \sim |x|^{-1/4}$ as $|x| \to \infty$. This is not sufficiently fast to make the integral converge.

Airy, Mathieu, and Weber beams, like their Bessel cousins, are self-healing after they encounter a small obstacle. This can be seen in figure 6.4, where a Weber beam recovers within a short distance after encountering the obstacle. The reader should be able to easily draw the pattern of rays, and see how the original profile is reconstructed by those rays that leak around the edge of the object.

6.3 Applications

A number of applications have been proposed or carried out with self-accelerating beams. For example, they have been used to trap and guide micro-particles [7, 8] and surface plasma paritons (acoustic excitations at the surfaces of solids) [9–11] along curved trajectories.

There is some evidence that Bessel, Airy, and other propagation-invariant beams may have benefits for microscopy [12] and other imaging methods. One example is in *optical coherence tomography* (OCT). OCT is a method of carrying out subsurface imaging of cross-sections in materials by interferometric means [13]. It is widely used in biology and medicine to view structures within organisms in a non-invasive manner. Besides having some potential benefits in terms of resolution and field of view, the self-healing property of such beams may allow viewing behind opaque sub-

Figure 6.4. The same mechanism that causes self-healing in Bessel beams works in Airy, Mathieu, and Weber beams as well. Here, a Weber beam, traveling left to right along a curved trajectory, encounters a small opaque sphere. After a short distance the disruption caused by the sphere has healed. (figure reproduced from [3].)

surface obstacles like hairs [14]. Airy beams have also been applied to light sheet microscopy [15], where single cross-sectional planes of a material are illuminated for viewing.

Bibliography

[1] Berry M V and Balazs N L 1979 Nonspreading wave packets *Am. J. Phys.* **47** 264
[2] Siviloglou G A, Broky J, Dogariu A and Christodoulides D N 2007 Observation of accelerating airy beams *Phys. Rev. Lett.* **99** 213901
[3] Zhang P, Hu Y, Li T, Cannan D, Yin X, Morandotti R C and Zhang X 2012 Nonparaxial Mathieu and Weber accelerating beams *Phys. Rev. Lett.* **109** 193901
[4] Sztul H I and Alfano R R 2008 The Poynting vector and angular momentum of Airy beams *Opt. Exp.* **16** 9411
[5] Chremmos I D and Efremidis N K 2013 Nonparaxial accelerating Bessel-like beams *Phys. Rev. A* **88** 063816
[6] Greenberger D M 1980 Comment on Nonspreading wave packets *Am. J. Phys.* **48** 256
[7] Baumgartl J, Mazilu M and Dholakia K 2008 *Nat. Photon.* **2** 675
[8] Zhang P, Prakash J, Zhang Z, Mills M S, Efremidis N K, Christodoulides D N and Chen Z 2011 Optically mediated particle clearing using Airy wavepackets *Opt. Lett.* **36** 2883
[9] Zhang P, Wang S, Liu Y, Yin X, Lu C, Chen Z and Zhang X 2011 Plasmonic Airy beams with dynamically controlled trajectories *Opt. Lett.* **36** 3191
[10] Minovich A, Klein A E, Janunts N, Pertsch T, Neshev D N and Kivshar Y S 2011 Generation and near-field imaging of Airy surface plasmons *Phys. Rev. Lett.* **107** 116802
[11] Li L, Li T, Wang S M, Zhang C and Zhu S N 2011 Plasmonic Airy beam generated by in-plane diffraction *Phys. Rev. Lett.* **107** 126804
[12] Fahrbach F O, Simon P and Rohrbach A 2010 Microscopy with self-reconstructing beams *Nat. Photon.* **4** 780
[13] Drexler W and Fujimoto J G (ed) 2008 *Optical Coherence Tomography–Technology and Applications* (Berlin: Springer)
[14] Blatter C, Grajciar B, Eigenwillig C M, Wieser W, Biedermann B R, Huber R and Leitgeb R A 2011 Extended focus high-speed swept source OCT with self-reconstructive illumination *Opt. Exp.* **19** 12141
[15] Vettenburg T, Dalgarno H I C, Nylk J, Coll-Lladó C, Ferrier D E K, Čižmár T, Gunn-Moore F J and Dholakia K 2014 Light-sheet microscopy using an Airy beam *Nat. Meth.* **11** 541

IOP Concise Physics

A Guided Tour of Light Beams
From lasers to optical knots
David S Simon

Chapter 7

Further variations

7.1 Separable solutions

A standard approach to solving many partial differential equations, including the Helmholtz equation, is to attempt separation of variables. This involves using trial solutions that factor into product form: each term in the product depends only on one coordinate and is independent of the others. For example, in Cartesian coordinates, the goal is to look for a solution of the form $E(x, y, z) = X(x)Y(y)Z(z)$. The three-dimensional equation will often then decouple into three one-dimensional equations. Such a method in spherical coordinates is often used to solve the Schrödinger equation for the hydrogen atom in introductory quantum mechanics textbooks [1, 2].

The form of the resulting separable solution will of course depend on the coordinate system used: a solution that is separable in Cartesian coordinates will not be separable in spherical coordinates, for example. Although cylindrical coordinates are the most common for describing optical beams, other possibilities exist and these lead to additional forms of separable solutions. For example, in Cartesian coordinates the separable solutions of the Helmholtz equation are the Hermite–Gauss modes, while in elliptical coordinates they are the Ince–Gauss modes. In cylindrical coordinates, the separable solutions are the Bessel modes. In total, the wave equation has been shown to be separable in 11 different orthogonal coordinate systems [3]. These allow an unambiguous separation between transverse and longitudinal directions, with translation invariance along the beam axis; these lead to the Whittaker-type diffraction-free solutions of section 5.3. A list of the fully separable solutions in these four coordinate systems is given in table 7.1. All of the solutions listed are diffraction-free. In addition, the Hermite–Gauss mode is separable in the transverse plane, but not in all three directions; as a result it does experience diffraction and beam-widening as it propagates.

Table 7.1. Fundamental separable beam solutions and the coordinate systems in which they separate. These are the only solutions that allow translation-invariant definition of transverse and longitudinal planes. These should be compared to the beam solutions that can be expressed in terms of Whittaker integrals; see table 5.1 in section 5.3. In addition, Hermite–Gauss beams are separable in the transverse plane.

Type of Solution	Coordinate system
Plane wave	Cartesian
Bessel	Circular cylindrical
Mathieu	Elliptical cylindrical
Parabolic	Parabolic cylindrical

In this chapter, we describe the most important beam solutions that are separable in two or three dimensions, and then discuss a couple of additional related topics such as Lorentz and elegant beams.

7.2 Hermite–Gauss beams

The Gaussian beam is the simplest type of beam that can be produced by a laser, but more complex modes are possible as well, with more complicated structure in the transverse spatial direction (transverse modes) or in the frequency dependence (longitudinal modes). The mode structure is determined by the geometry of the laser's resonant cavity. Longitudinal modes are determined by the length of the cavity in the propagation direction, while transverse modes are determined by the shape of the mirrors at the end of the cavity and by the shape of the cavity itself in a transverse plane. Most commonly, the mirrors are spherical, leading to Gaussian or higher TEM (transverse electromagnetic) modes.

Consider longitudinal modes first. Let L be the length of the cavity and n the refractive index of the material in the cavity. Resonant modes in the cavity will have nodes at the mirrors, so that the allowed wavelengths are those satisfying

$$L = \frac{m\lambda}{2} = \frac{m\pi c}{n\omega}, \qquad (7.1)$$

since $\lambda = \frac{c}{f} = \frac{2\pi c}{\omega}$. Here m can be any positive integer. Solving for the angular frequency, the allowed spectrum consists of a set of equally spaced frequencies

$$\omega_m = \frac{m\pi c}{nL}, \qquad (7.2)$$

separated by distance

$$\Delta\omega = \frac{\pi c}{nL}. \qquad (7.3)$$

The cavity length is not the whole story, of course: the gain material in the laser cavity absorbs and emits photons during electron energy-level transitions within the atoms. As discussed in chapter 3, these atomic transitions have an intrinsic spectral line-shape $g(\omega)$, defined as the photon emission probability of the atom per unit frequency. This line-shape is peaked at the resonant frequency of the transition, must obey the normalization condition $\int g(\omega)d\omega = 1$, and is typically Lorentzian in shape. Several of the resonant frequencies of the cavity may be included within the region where $g(\omega)$ is of significant size (figure 3.3). All of these longitudinal modes will in general occur, leading to a beam with multiple frequencies.

But having multiple frequencies is bad for the quality of the resulting beam, since the modes will interfere, leading to beat frequencies and loss of coherence. Therefore, some mechanism is usually introduced to block all but one of these frequency modes. One common method for doing so is the use of a Fabry–Perot etalon [4], which passes a narrower frequency band, leaving just a single sharply-peaked spectral line in the output.

Turning to transverse modes, the outgoing beam may have non-trivial spatial structure in the plane perpendicular to the propagation axis. This structure may include position-dependent variations of both amplitude and phase. If a sufficiently small aperture is placed in the cavity, only the Gaussian beam will survive. But without such an aperture, additional modes will appear. The result will be *Hermite–Gauss* or TEM modes. These are specified by two integers m and n, and are denoted as $TM_{m,n}$ modes. The complex field amplitude of the $T_{m,n}$ mode is given in Cartesian coordinates by:

$$E_{m,n}(x,y) = E_0 \left(\frac{w_0}{w(z)}\right) H_m\left(\frac{\sqrt{2}x}{w(z)}\right) H_n\left(\frac{\sqrt{2}y}{w(z)}\right) \quad (7.4)$$
$$\times e^{-(x^2+y^2)/w^2} e^{-ikz-ik(x^2+y^2)/2R(z)+i(m+n+1)\zeta(z)},$$

where w and R are the waist size and radius of curvature, ζ is the Guoy phase, and H_m denotes the mth Hermite polynomial. Recall that the first few Hermite polynomials are given by

$$H_0(u) = 1 \qquad H_2(u) = 4u^2 - 2 \quad (7.5)$$

$$H_1(u) = 2u \qquad H_3(u) = 8u^2 - 12u. \quad (7.6)$$

The Hermite–Gauss modes have a rectangular symmetry, as shown in figure 7.1; they are separable functions in the x–y plane. The mode TM_{mn} has $m+1$ maxima in the x direction and $n+1$ in the y direction. The lowest order Hermite–Gauss mode, TM_{00}, is simply the Gaussian discussed in section 3.2. Notice, that although these solutions are separable in x and y, they are not separable in z; the z variable enters into the same Hermite factors as the other two coordinates, not in a separate factor of its own. As a result the Hermite–Gauss beams are not translation-invariant along the z axis, and will display dispersive effects like beam divergence.

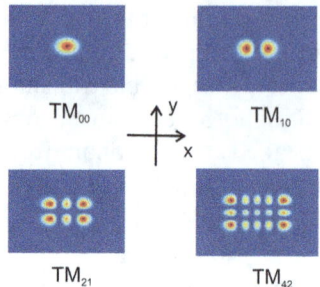

Figure 7.1. Intensity cross-sections of Hermite–Gauss modes. A TM_{mn} mode has $m+1$ maxima in the x direction and $n+1$ in the y direction.

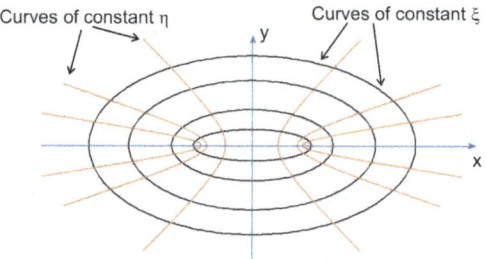

Figure 7.2. Elliptical coordinates on the plane. ξ serves as a radial variable, and the ellipses in this figure are curves of constant ξ. The parabolas are curves of constant η; η serves roughly as an angular variable. This system reduces to standard polar coordinates as the ellipticity ϵ goes to zero, and reduces to Cartesian coordinates as $\epsilon \to \infty$.

7.3 Ince–Gauss beams

Starting from Cartesian coordinates (x, y, z), an elliptical coordinate system (ξ, η, z) may be defined according to the relations

$$x = \epsilon \cosh \xi \cos \eta \tag{7.7}$$

$$y = \epsilon \sinh \xi \sin \eta, \tag{7.8}$$

where $\xi \in \{0, \infty\}$ and $\eta \in \{0, 2\pi\}$ play roles analogous to radial and angular variables, respectively. The dimensionless parameter ϵ is the ellipticity parameter. Curves of constant ξ are ellipses, while curves of constant η are parabolas (figure 7.2). Cylindrical coordinates can be viewed as a special case of elliptical coordinates, in which $\epsilon \to 0$. Similarly, Cartesian coordinates are obtained from elliptical coordinates by $\epsilon \to \infty$.

The Helmholtz equation in elliptical coordinates takes the form

$$\frac{1}{f^2(\cosh^2 \xi - \cos^2 \eta)} \left(\frac{\partial E(r)}{\partial \xi^2} + \frac{\partial E(r)}{\partial \eta^2} \right) + 2ik \frac{\partial E(r)}{\partial z} = 0, \tag{7.9}$$

and the separable solutions of this equation are the *Ince–Gauss modes* [5]. Here, $f(z) = f_0 \sqrt{1 + (z/z_R)^2}$, z_R is the Rayleigh range, and f_0 is related to the ellipticity parameter by $\epsilon = 2(f_0/w_0)^2$. The equation is solved by seeking separable solutions of the form

$$E(r) = G(\xi)N(\eta)e^{iZ(z)}\Psi(r), \qquad (7.10)$$

where $\Psi(r)$ is a Gaussian beam. The Helmholtz equation then separates into three independent equations,

$$\frac{d^2G}{d\xi^2} - \epsilon \sinh(2\xi)\frac{dG}{d\xi} - (a - p\epsilon \cosh(2\xi))G = 0 \qquad (7.11)$$

$$\frac{d^2N}{d\eta^2} + \epsilon \sin(2\eta)\frac{dN}{d\eta} + (a - p\epsilon \cos(2\eta))N = 0 \qquad (7.12)$$

$$-\left(\frac{z^2 + z_R^2}{z_R}\right)\frac{dZ}{dz} = p, \qquad (7.13)$$

where a and p are separation constants.

The solutions that result are of the form

$$E_{p,m}^e = D\frac{w_0}{w(z)}C_p^m(i\xi, \epsilon)C_p^m(\eta, \epsilon) \\ \times e^{-r^2/w^2(z)}e^{ikz+i\frac{\xi kr^2}{2R(z)}-i(p+1)\arctan(z/z_R)}, \qquad (7.14)$$

$$E_{p,m}^o = D\frac{w_0}{w(z)}S_p^m(i\xi, \epsilon)S_p^m(\eta, \epsilon) \\ \times e^{-r^2/w^2(z)}e^{ikz+i\frac{\xi kr^2}{2R(z)}-i(p+1)\arctan(z/z_R)}, \qquad (7.15)$$

where D is a constant, $w(z)$ and w_0 are the beam radius at z and at the waist. z_R and $R(z)$ are the Rayleigh range and the radius of curvature. The functions C_p^m and S_p^m are the even and odd Ince polynomials [6] of order p and degree m. Examples of phase and amplitude profiles of Ince–Gauss beams are shown in figure 7.3.

7.4 Parabolic beams

The remaining class of fully separable, non-diffracting beams is the set of *parabolic beams* [7], which are separable functions in the parabolic cylindrical coordinate system $\{\eta, \xi, z\}$. η and ξ are related to the transverse Cartesian coordinates x and y by

$$x + iy = (\eta + i\xi)^2/2. \qquad (7.16)$$

η ranges from $-\infty$ to $+\infty$, while the coordinate ξ is non-negative, $0 \leq \xi \leq \infty$ (figure 7.4). In these coordinates, the Helmholtz equation decouples into three independent equations, leading to a factorable field solution, $E(r) = Z(z)\Phi(\eta)R(\xi)$.

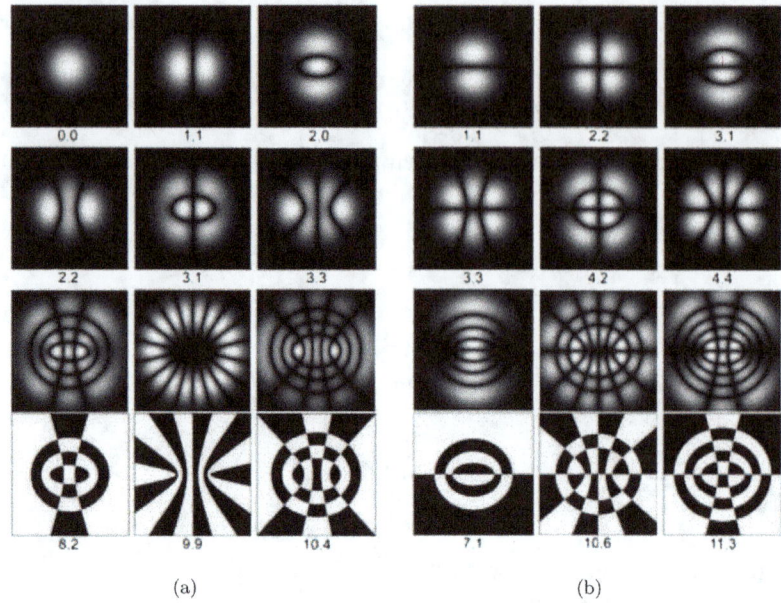

Figure 7.3. Examples of the transverse-plane amplitude (top three rows) and phase (bottom row) profiles of Ince–Gauss beams. (a) shows even beams, where (b) shows odd. The last row shows the phase profile of the row immediately above it. (Figures reproduced from [5].)

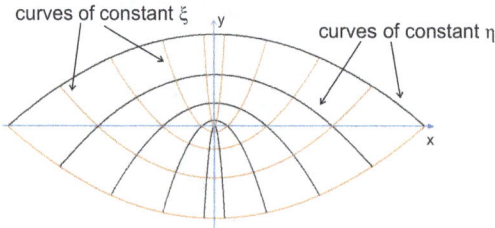

Figure 7.4. Parabolic coordinate on the plane. The red curves are curves of constant ξ, and the black ones are curves of constant η. Note that the constant η and constant ξ curves are interchanged by the reflection $y \to -y$.

The z equation gives the usual e^{-ikz} propagation factor. The remaining two transverse equations take the form

$$\frac{d^2\Phi(\eta)}{d\eta^2} + \left(k_\perp^2\eta^2 + 2k_\perp a\right)\Phi(\eta) = 0 \tag{7.17}$$

$$\frac{d^2 R(\xi)}{d\xi^2} + \left(k_\perp^2\xi^2 - 2k_\perp a\right)R(\xi) = 0, \tag{7.18}$$

where $k_\perp$ is the transverse momentum and the seperation constant is written as $2k_\perp a$. Each of these equations can be transformed into the standard form for parabolic cylinder differential equations (see [7] for details) and solved in the form of a Taylor series expansion with coefficients related by a recurrence relation. Like the

Ince–Gauss modes, the transverse parabolic solutions at any fixed z divide into even and odd types. These are given by

$$E_e(\eta, \xi, a) = \frac{1}{\pi\sqrt{2}} |\Gamma_1|^2 P_e(\xi\sqrt{2k_\perp}, a) P_e(\eta\sqrt{2k_\perp}, -a) \qquad (7.19)$$

$$E_o(\eta, \xi, a) = \frac{\sqrt{2}}{\pi} |\Gamma_3|^2 P_o(\xi\sqrt{2k_\perp}, a) P_o(\eta\sqrt{2k_\perp}, -a), \qquad (7.20)$$

where $\Gamma_1 = \Gamma(\frac{1}{4} + \frac{ia}{2})$ and $\Gamma_3 = \Gamma(\frac{3}{4} + \frac{ia}{2})$, Γ is the Euler gamma function, and the P functions are given by

$$P(v, a) = \sum_{n=0}^{\infty} c_n \frac{v^n}{n!}, \qquad (7.21)$$

where the coefficients obey the recurrence relation

$$c_{n+2} = ac_n - \frac{n(n-1)}{4} c_{n-2}. \qquad (7.22)$$

The difference between the even and odd functions is that the initial conditions for P_e are $c_0 = 1$, $c_1 = 0$, while for P_o these are given by $c_0 = 1$, $c_1 = 0$. The result is a solution that falls into the Whittaker form (equation (5.7)), with the angular spectral functions for the even and odd cases given by:

$$f_e(\phi) = \frac{1}{2(\pi|\sin\phi|)^{1/2}} e^{ia \ln |\tan \phi/2|} \qquad (7.23)$$

$$f_o(\phi) = \begin{cases} if_e(\phi), & \text{if } \phi \in (-\pi, 0) \\ -if_e(\phi), & \text{if } \phi \in (0, \pi). \end{cases} \qquad (7.24)$$

Examples of even and odd parabolic beams are shown in figure 7.5.

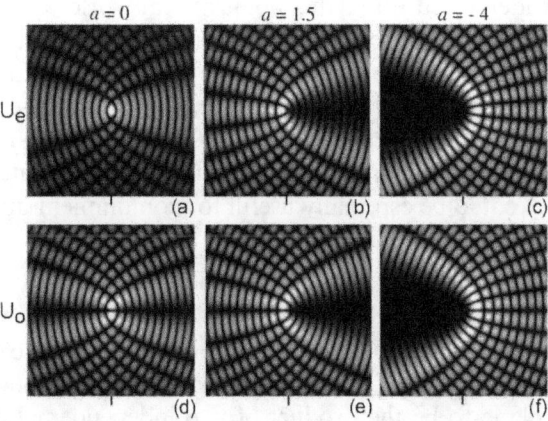

Figure 7.5. Transverse cross-sections of parabolic beams. The top row gives the amplitudes of even beams for different values of parameter a, the bottom row has the same for odd beams. (Figures reproduced from [7].)

A non-paraxial, self-accelerating variation on the parabolic beam exists [8], which exactly solves the paraxial Helmholtz equation in parabolic coordinates. This solution is called a *Weber beam*.

7.5 Elegant beams

Recall that in section 7.2 the Hermite–Gauss modes were described by the equation

$$E_{m,n}(x, y) = E_0 \left(\frac{w_0}{w(z)}\right) H_m\left(\frac{\sqrt{2}x}{w(z)}\right) H_n\left(\frac{\sqrt{2}y}{w(z)}\right) \quad (7.25)$$
$$\times e^{-(x^2+y^2)/w^2} e^{-ikz-ik(x^2+y^2)/2R(z)+i(m+n+1)\zeta(z)}.$$

Notice that the factors $\frac{x}{w}$ and $\frac{y}{w}$ occur both inside the Hermite polynomials and in the exponent. But in one case the they are multiplied by a real number, while in the other they are multiplied by the imaginary number i. It has been noted that if the Hermite polynomial factors $H_m(\frac{\sqrt{2}x}{w(z)})$ $H_n(\frac{\sqrt{2}y}{w(z)})$ are complexified by inserting a factor of i into their arguments, $H_m(\frac{\sqrt{2}x}{w(z)}) \to H_m(\frac{i\sqrt{2}x}{w(z)})$, then the resulting expression is still a solution to the Helmholtz equation but is much more mathematically elegant, with simplifications occurring in its use. As a result, this complexified version is called the *'elegant' Hermite–Gauss beam* [9, 10] $\widetilde{E}_{m,n}$. The elegant Hermite–Gauss functions still constitute a mathematically complete set. However, they are now orthogonal only under an altered definition of the inner product. Whereas the usual Hermite–Gauss functions (proportional to $H_n(\frac{\sqrt{2}x}{w(z)})$) are orthogonal to each other, the elegant case (proportional to $H_n(\frac{\sqrt{2}ix}{w(z)})$) obeys an orthogonality relation with the dual or adjoint function $v(x) = H_n(-\frac{\sqrt{2}ix}{w(z)})$ [9, 10]. Such functions, which are not orthogonal among themselves but instead orthogonal to a set of adjoint functions, are called *biorthogonal*.

Although less widely used up to this point in real-world applications than their standard counterparts, elegant solutions of other beam types have been considered, such as the elegant Laguerre–Gauss [11, 12], elegant Hermite–Laguerre–Gauss [13] and elegant Ince–Gauss beams [14]. It has been recently pointed out [15] that the elegant Laguerre–Gauss and elegant Hermite–Gauss beams can produce stronger confining forces on dielectric particles than their standard counterparts, raising the possibility that they could be especially useful for containment and manipulation of nanoparticles.

7.6 Lorentz beams

A further type of directed optical beam has been studied. These are called *Lorentz beams* [16], because they are produced by arranging for the amplitude in the transverse source plane to be the product of two independent Lorentzian distribution functions. Recall that a normalized Lorentzian function is of the form

$$L(x) = \frac{\Gamma/2\pi}{(x-x_0)^2 + (\Gamma/2)^2} = \frac{(2/\pi\Gamma)}{1 + \dfrac{(x-x_0)^2}{(\Gamma/2)^2}}, \qquad (7.26)$$

where Γ is a measure of the width of the Lorentzian distribution. The Lorentzian distribution function is also sometimes called the *Breit–Wigner distribution* or the *Cauchy distribution*, and appears in the study of high-energy resonances in nuclear physics and of the thermal broadening of spectral lines.

The amplitude of a Lorentz beam in the source plane is of the form

$$E(x, y) = \frac{C}{\left(1 + (x/w_x)^2\right)\left(1 + (y/w_y)^2\right)}, \qquad (7.27)$$

where C is a constant, and w_x and w_y are the widths in the x and y directions. From the amplitude distribution in the initial plane an analytic expression for the transverse amplitude at any distance z can be analytically calculated [16]. Adding an azimuthally varying phase factor, these beams can be given orbital angular momentum. And with an extra Gaussian modulation $e^{-(x^2+y^2)/w^2}$ in the initial plane they can be made into *Lorentz–Gauss beams*. Lorentz–Gauss beams have a Lorentzian profile near the axis and Gaussian profile farther out. Although proposed relatively recently, these beams have already been well-studied in a number of different contexts; see for example [17–21].

The Lorentz beams are of interest because they provide links between concrete optical structures and several types of abstract mathematical structures. For example, their properties can be derived by means of Lie group theory [16], a method which makes the symmetry properties of the beams explicit. Similarly, it has been shown [22] that the Lorentz beams have a close mathematical connection to the so-called *relativistic Hermite polynomials* [23–29]. These relativistic polynomials have been used to study quantum mechanical harmonic oscillators in the relativistic limit, which can be useful for modeling the behavior of quarks inside a nucleus or of quantum fields in curved space-time. The relativistic Hermite polynomials reduce to the ordinary Hermite polynomials in the low-speed limit; similarly, the Hermite–Gauss and elegant Hermite–Gauss beams can be obtained from the Lorentz–Gauss beams in the analogous limit [22].

Bibliography

[1] Shankar R 1994 *Principles of Quantum Mechanics* 2nd edn (New York: Plenum)
[2] Griffiths D J 2004 *Introduction to Quantum Mechanics* 2nd edn (Upper Saddle River, NJ: Pearson-Prentice Hall)
[3] Miller W Jr 1977 Symmetry and separation of variables *Encyclopedia of Mathematics and its Applications* vol 4 (Reading, MA: Addison-Wesley)
[4] Saleh B E A and Teich M C 2007 *Fundamentals of Photonics* 2nd edn (Hoboken, NJ: Wiley)
[5] Bandres M A and Gutiérrez-Vega J C 2004 Ince-Gaussian beams *Opt. Lett.* **29** 144
[6] Arscott F M 1964 *Periodic Differential equations* (Oxford: Pergamon Press)

[7] Bandres M A, Gutiérrez-Vega J C and Chávez-Cerda S 2004 Parabolic nondiffracting optical wave fields *Opt. Lett.* **29** 44

[8] Zhang P, Hu Y, Li T, Cannan D, Yin X, Morandotti R C and Zhang X 2012 Nonparaxial Mathieu and Weber accelerating beams *Phys. Rev. Lett.* **109** 193901

[9] Siegman A E 1973 Hermite-Gaussian functions of complex argument as optical-beam eigenfunctions *J. Opt. Soc. Am.* **63** 1093

[10] Siegman A E 1986 *Lasers* (Mill Valley, CA: University Science Books)

[11] Zauderer E 1986 Complex argument Hermite-Gaussian and Laguerre–Gaussian beams *J. Opt. Soc. Am.* A **3** 465

[12] Saghafi S and Sheppard C J R 1998 The beam propagation factor for higher order Gaussian beams *Opt. Commun.* **153** 207

[13] Deng D and Guo Q 2008 Elegant Hermite-Laguerre–Gaussian beams *Opt. Lett.* **33** 1225

[14] Bandres M A 2004 Elegant Ince-Gaussian beams *Opt. Lett.* **29** 1724

[15] Alpmann C, Schöler C and Denz C 2015 Elegant Gaussian beams for enhanced optical manipulation *Appl. Phys. Lett.* **106** 241102

[16] El Gawhary O and Severini S 2006 Lorentz beams and symmetry properties in paraxial optics *J. Opt.* A **8** 409

[17] Ni Y and Zhou G 2013 Propagation of a Lorentz-Gauss vortex beam through a paraxial ABCD optical system *Opt. Commun.* **291** 19

[18] Zhoung G, Wang X and Chu X 2013 Fractional Fourier transform of Lorentz-Gauss vortex beams *Sci. Chin. Phys. Mech. Astron.* **56** 1487

[19] Zhoung G and Ru G 2013 Propagation of Lorentz-Gauss vortex beam in a turbulent atmosphere *PIER* **143** 143

[20] Ni Y and Zhou G 2012 Nonparaxial propagation of Lorentz–Gauss vortex beams in uniaxial crystals orthogonal to the optical axis *Appl. Phys.* B **108** 883

[21] Torre A 2016 Wigner distribution function of a Lorentz-Gauss vortex beam: alternative approach *Appl. Phys.* B **122** 55

[22] Torre A, Evans W A B, El Gawhary O and Severini S 2008 Relativistic Hermite polynomials and Lorentz beams *J. Opt.* A **10** 115007

[23] Aldaya V, Bisquert J and Navarro-Salas J 1991 The quantum relativistic harmonic oscillator: generalized Hermite polynomials *Phys. Lett.* A **156** 381

[24] Nagel B 1994 The relativistic Hermite polynomial is a Gegenbauer polynomial *J. Math. Phys.* **35** 1549

[25] Dattoli G, Lorenzutta S, Maino G and Torre A 1998 The generating function method and properties of the relativistic Hermite polynomials *Nuovo Cimento* B **113** 553

[26] Ismail M E H 1996 Relativistic orthogonal polynomials are Jacobi polynomials *J. Phys.* A **29** 3199

[27] Navarro D J and Navarro-Salas J 1997 Special-relativistic harmonic oscillator modeled by Klein–Gordon theory in anti-de Sitter space *J. Math. Phys.* **37** 6060

[28] Simon D S 2011 Supersymmetry and duality in relativistic oscillator systems *Phys. Lett.* A **375** 3751

[29] Baskal S, Kim Y S and Noz M E 2015 *Physics of the Lorentz Group* (New York: Morgan and Claypool Publishers)

IOP Concise Physics

A Guided Tour of Light Beams
From lasers to optical knots
David S Simon

Chapter 8

Entangled beams

8.1 Separability and entanglement

This chapter focuses on quantum entangled beams, in which two beams have linked properties. Such entangled beams have become useful in a number of areas in recent years. In this chapter, the word 'beam' will be stretched a bit; in many cases the beam in question will consist of only a single photon at a time.

In quantum mechanics, an individual particle such as an electron or a photon is described by a wavefunction, $\psi(x)$, or equivalently by a state vector $|\psi\rangle$. So it is natural to assume that when two particles are present, a wavefunction may be written for each of them, and then the total wavefunction of the two-particle system is simply the product of the two: in other words, the amplitude of finding particle a at location x_1 and simultaneously finding particle b at x_2 should be

$$\psi(x_1, x_2) = \psi_a(x_1)\psi_b(x_2). \tag{8.1}$$

If the two particles are distinguishable (a photon and an electron, for example, or a vertically-polarized photon and a horizontally-polarized photon) then this is correct. However, what if the two particles are indistinguishable, such as two vertically-polarized photons of the same frequency propagating along the same direction? No measurement will be able to tell you if it is photon a at point x_1 and photon b at x_2, or vice versa. Therefore, the superposition principle of quantum mechanics tells us that both possibilities must be added. The full wavefunction of the two-photon system is then

$$\psi(x_1, x_2) = \frac{1}{\sqrt{2}}\big(\psi_a(x_1)\psi_b(x_2) + \psi_a(x_2)\psi_b(x_1)\big). \tag{8.2}$$

The factor of $\frac{1}{\sqrt{2}}$ is there to keep the composite state properly normalized, and the relative plus sign between the terms is due to the bosonic nature of the photons, which requires the wavefunction to be symmetric under particle interchange.

(The two-electron state has a relative minus sign, signaling the antisymmetry required of fermions.)

Suppose now that the two photons have different polarizations or different frequencies. Then appropriate measurements can distinguish between them. But the apparatus used in a given experiment may not allow such measurements, in which case we will again find a non-factorable wavefunction like equation (8.2). For example, two photons of opposite polarization (vertical and horizontal) may be produced together and sent through a (non-polarizing) beam splitter. After the beam splitter, imagine that one photon is transmitted, leading to its detection at point x_1, while the other is reflected to a detector at point x_2. The polarizations of the photons are never measured at any point. As a result it is impossible to know which photon took which path, so both possibilities must again be added. The state vector just before detection is therefore of the form

$$|\psi\rangle = \frac{1}{\sqrt{2}}\left(|H\rangle_{x_1}|V\rangle_{x_2} + |H\rangle_{x_2}|V\rangle_{x_1}\right), \tag{8.3}$$

where, for example, $|H\rangle_{x_j}$ means a horizontally-polarized photon at x_j, and similarly for vertically polarized $|V\rangle_{x_j}$. Once again it is not possible to factor the two-particle wavefunction into separate single-particle wavefunctions for the two photons, the state again has the same form as equation (8.2). If the polarization of one photon is measured, then the act of measurement reduces the wavefunction of equation (8.3) to a single term; for example, if the photon at x_1 is measured to be vertically-polarized, then the wavefunction is reduced to $|\psi\rangle = |H\rangle_{x_2}|V\rangle_{x_1}$. But until such a measurement is made, the wavefunction will remain in the two-term superposition of equation (8.3).

A two-particle superposition state (or more generally any n-particle state) like equation (8.2) or equation (8.3) is called *entangled*. An entangled state is one that cannot be factored into separate single-particle states for each particle in the composite system. A state such as that of equation (8.1), which can be factored, is called *separable*. Particles in a separable state are essentially independent: a measurement of the wavefunction of particle a tells you nothing about the state of particle b. Particles in an entangled state, on the other hand, are intrinsically linked with each other in some sense: if the photon at x_1 in the state of equation (8.3) is measured and found to have vertical polarization, then it is automatically known that the photon at x_2 is horizontally-polarized. The properties of the two entangled photons are therefore perfectly correlated or anti-correlated with each other: when one is found to be horizontal, the other has to be vertical.

Two things should be noticed about the correlations between entangled particles. First, this quantum entanglement differs fundamentally from the correlations found in classical systems. For example, suppose two photons of opposite polarization (horizontal and vertical) are produced with classical anti-correlation between the particles. This means that the polarization of each photon is random (either H or V), but that when one is H the other is always found to be V. For a particular pair, suppose photon a is horizontal and b is vertical; the pair has separable wavefunction

$|\psi\rangle = |H\rangle_a|V\rangle_b$. But what happens if the polarizations are measured along diagonal axes, at 45° to the horizontal and vertical? In this case, photon a has equal probability of being found to have polarization along each diagonal. The same is true of photon b. More importantly, there is no correlation between the polarization of a and the polarization of b in this diagonal system! Each photon projects onto the diagonal axes in a manner that is independent of the other. Thus, the correlation that was present between the polarizations in the horizontal–vertical coordinate system vanishes in the rotated, diagonal system. This is characteristic of classical correlations: they are apparent in the coordinate system in which the system was prepared, but vanish when other types of measurements are made.

In contrast, suppose the entangled state

$$|\psi\rangle = \frac{1}{\sqrt{2}}(|H\rangle_a|V\rangle_b + |H\rangle_b|V\rangle_a) \tag{8.4}$$

is created instead. The particles are again prepared in such a way as to be correlated in the horizontal–vertical coordinate system. But now, the correlation will survive rotation to any other system. For example, suppose that the coordinate system is again rotated by 45° degrees to give two new axes D (diagonal) and A (antidiagonal). It should be straightforward for the reader to carry out the rotation using

$$\begin{pmatrix}|D\rangle \\ |A\rangle\end{pmatrix} = \begin{pmatrix}\cos\theta & -\sin\theta \\ \sin\theta & \cos\theta\end{pmatrix}\begin{pmatrix}|H\rangle \\ |V\rangle\end{pmatrix}, \tag{8.5}$$

and to thereby verify that the wavefunction in this new system is given by

$$|\psi\rangle = \frac{1}{\sqrt{2}}(|D\rangle_a|A\rangle_b + |A\rangle_b|D\rangle_a). \tag{8.6}$$

The wavefunction still has the same form in the rotated coordinate system; in particular, the polarizations are still anticorrelated! This will remain true in any other rotated system as well.

So the quantum correlations of the entangled system are in a sense stronger than any possible classical correlation, since they survive changes in the choice of how the measurements are done. This is related to the fact that in classical mechanics a variable (polarization in this case) always has a definite value, whether that value is known or not. In contrast, quantum mechanical variables may be in superposition of several values (or even infinitely many values) at the same time. The value only becomes definite when the variable is measured, or alternatively when interactions with the surroundings have reduced the system to a classical state (a process known as *decoherence*). Until then, the variables describing the system have no definite well-defined values, but instead have a set of well-defined *amplitudes* for all of their possible values. The excess strength of the quantum correlations comes from the fact that *all* of these amplitudes must be anti-correlated in a consistent manner. These facts are given a more quantitative meaning through the well-known Bell and CHSH inequalities [1–5], which were developed in response to the Einstein–Podolski–Rosen (EPR) 'paradox'. A discussion of these matters would take us too far afield here, but introductions to these inequalities at various levels of sophistication may be found in many places, including [6–10].

The second thing to notice about the correlations in entangled systems are that they are non-local. No matter how far apart the two photons travel, the polarizations will be found to be orthogonal to each other, regardless of which measurement axes are used. But some thought should make it clear that there is no way to use this correlation to superluminally transmit information between the two measurement sites. At each end, the measurement outcomes are perfectly random; in order for any useful correlations to be visible, the results of the measurements have to be brought back together for comparison. This of course can only be done at subluminal speeds.

Entangled light beams have found many applications in recent years, and have become one of the chief tools in the area of quantum optics. In the next section, the most common means of generating entangled photons will be discussed, and then in the following sections a few applications will be mentioned. More detailed reviews of the creation and applications of entangled light beams may be found in [11–13].

8.2 Creating entanglement

In principle there are many ways of entangling photons. But in practice, most experiments using entangled photons make use of the process of *spontaneous parametric down conversion* (SPDC), also known as *parametric fluorescence*. This is a phenomenon first discovered in microwave engineering, and then later applied to optical systems.

When the electromagnetic field in an optical wave is small enough, the wave and the propagation medium interact in a linear manner, with the interaction strength being proportional to the field strength. However, when the field strength increases, nonlinear interactions become more important. Many readers may be familiar with similar situations in other areas of physics. For example, in mechanics when a particle is moving with small displacements around the minimum of a potential, then regardless of the overall form of the potential, its Taylor series may always be truncated to keep just the quadratic terms. For small amplitudes, the potential can therefore always be approximated by a harmonic oscillator potential. However, if the particle is given more energy, it will start moving farther and farther away from the minimum, and will begin exploring regions where the quadratic approximation breaks down. Thus, for example, springs that are driven at large amplitudes will no longer obey Hooke's law and the spring becomes nonlinear.

Consider light passing through a crystalline material. The oscillating electric field of the light will periodically distort the molecules in the material as the wave propagates. This distortion causes positively and negatively charged portions of the molecule to be displaced from their equilibrium positions: in other words, the molecule develops a periodically oscillating electric dipole moment. The dipole moment per unit volume of the material is called the *polarization*. Linear materials are those in which it is safe to model the polarization as being proportional to the passing electric field: $\boldsymbol{P}(\boldsymbol{r}, t) = \epsilon_0 \chi \boldsymbol{E}(\boldsymbol{r}, t)$. Here, χ is the *electric susceptibility* of the material, which is related to the index of refraction by $n = \sqrt{1 + \chi}$.

However, in nonlinear materials, this approximation cannot be made at reasonable field strengths. Higher order terms in the polarization may not be ignored, so we must expand:

$$P_i(\boldsymbol{r}, t) = \epsilon_0 \left(\sum_j \chi_{ij}^{(1)} E^j + \sum_{jk} \chi_{ijk}^{(2)} E^j E^k + \sum_{jkl} \chi_{ijkl}^{(3)} E^j E^k E^l + \cdots \right). \qquad (8.7)$$

Here, the indices $j, k, l,...$ indicate spatial components, and we make no distinction between upper and lower indices. Note that in general the susceptibilities are tensor quantities. For our purposes, the relevant term is the one involving second-order susceptibility, $\chi^{(2)}$; we will ignore the other terms. The energy associated with this term is then

$$H = \int d^3 r \, \boldsymbol{P} \cdot \boldsymbol{E} = \sum_l P_l E^l = \sum_{jkl} \chi_{jk} E^j E^k E^l. \qquad (8.8)$$

(We denote energy by H (for Hamiltonian), rather than E, so that there is no confusion with the electric field.) A quadratic polarization therefore leads to an energy with a cubic field dependence.

Anyone who has studied high-energy physics is familiar with the idea that a term in a Hamiltonian that is nth-order in a field represents an interaction term between n particles. Specifically, a cubic Hamiltonian like the one above describes a three-particle interaction, involving three fundamental quanta of the relevant field. In our case, we deal with electromagnetic fields so that the fundamental quantum involved is the photon. The cubic energy term therefore represents a three-photon interaction. This may seem odd, since photons have no electric charge, and therefore do not interact with each other (at least not electromagnetically). The interaction is indirect; in this case it is mediated by the nonlinear crystal. Each photon interacts with the same crystal lattice, which acts to transfer energy between them. Effectively, the lattice is acting like a nonlinear spring connected between the photons. The crystal is also necessary here because energy and momentum conservation cannot be simultaneously satisfied for such interactions in vacuum; the crystal lattice carries off any excess energy and momentum, making the interaction possible.

Although the energy is cubic in the field, the polarization is quadratic, so this interaction is usually called a second-order nonlinear interaction. Second-order interactions in the crystal can take several forms. For example two low energy photons can annihilate to produce a new higher-energy photon. This is called *second harmonic generation*. In the present case, we are interested in the inverse of this process: an incident high-energy photon (usually ultraviolet) called the *pump photon* enters the crystal with energy $E_p = \hbar \omega_p$ and splits into two lower-energy photons, called for historical reasons the signal and idler photons. These two daughter photons (often of visible or infrared frequencies) have energies $E_s = \hbar \omega_s$

and $E_i = \hbar\omega_i$. Since energy and momentum must be conserved, the following relations (the *phase-matching conditions*) must be obeyed by the frequencies and wavevectors:

$$\omega_p = \omega_s + \omega_i \tag{8.9}$$

$$\boldsymbol{k}_p = \boldsymbol{k}_s + \boldsymbol{k}_i. \tag{8.10}$$

These conditions can also be viewed as constructive interference conditions between the three photons.

The pump photons are usually provided by a laser. Since only a small fraction of the pump photons undergo down conversion (often one in a million or less), the leftover pump beam usually has to be removed. This can be done using a *spectral filter* that blocks the pump portion of the spectrum but allows the lower frequency signal and idler to pass. The same result can be obtained by means of a *dichroic mirror*, which reflects short wavelengths and transmits long wavelengths. By controlling the properties of the pump laser and the crystal, the properties of the down converted pairs can be varied over a wide range.

The photons produced by SPDC are highly entangled with each other. In fact they are *hyperentangled*, having strong quantum correlation or anticorrelation simultaneously in multiple degrees of freedom: frequency, momentum, polarization, and orbital angular momentum. These strong entanglement properties and the high level of controllability make down conversion a very versatile tool and the resulting signal/idler pairs ideal for many types of quantum mechanical experiments such as tests of Bell inequalities. As will be seen below, they have other applications as well.

The frequency (or energy) of the signal and idler are always anticorrelated: if $\omega_s = \omega_0 + \Omega$ for some Ω, then $\omega_i = \omega_0 - \Omega$, where $\omega_0 = \frac{\omega_p}{2}$. Similarly, if the pump has no orbital angular momentum, then the signal and idler angular momenta about the propagation axis must be equal and opposite: $L_s = l\hbar$ and $L_i = -l\hbar$. If we ignore all other variables, the angular momentum part of the outgoing signal/idler state is of the form

$$|\Psi\rangle = \sum_l C_l |l\rangle_s |-l\rangle_i, \tag{8.11}$$

for some set of coefficients C_l, where $|\pm l\rangle$ represents a state of definite OAM $\pm l$ (a Laguerre–Gauss state, for example). This shows the entanglement of the state in terms of OAM eigenstates.

In the same way, the outgoing two-photon state may be entangled in polarization. There are two cases. In *type I* down conversion, the signal and idler have the same polarization. In this case, the polarization part of the wavefunction is in general of the entangled form

$$|\Psi\rangle = \frac{1}{\sqrt{2}}(|H\rangle_s |H\rangle_i + |V\rangle_s |V\rangle_i). \tag{8.12}$$

In *type II* down conversion the signal and idler have opposite polarization, leading to a different polarization-entangled output state:

$$|\Psi\rangle = \frac{1}{\sqrt{2}}(|H\rangle_s|V\rangle_i + |V\rangle_s|H\rangle_i). \tag{8.13}$$

In experiments where entanglement plays a role, it is common to use pump beams that have been strongly attenuated, so that the probability is very low of having more than one entangled output pair present in the experiment at a given time. The goal is to measure correlations between the two photons. This is done by means of *coincidence detection*: two detectors are used, and a count is recorded only when photons reach both detectors within a very short time window. These coincidence counts then provide a measurement of the *second-order correlation functions* (or intensity correlation functions) in the system,

$$g^{(2)}(\tau) = \langle I(t+\tau)I(t)\rangle = \langle E^\dagger(t+\tau)E^\dagger(t)E(t+\tau)E(t)\rangle, \tag{8.14}$$

where $I(t) = |E(t)|^2$ is the intensity and the brackets $\langle ...\rangle$ represent averages or expectation values. This is in contrast with single-detector setups, which can only measure the first-order correlation functions of the field (also called amplitude correlation functions),

$$g^{(2)}(\tau) = \langle E^\dagger(t+\tau)E(t)\rangle. \tag{8.15}$$

Many quantum effects that are invisible to first-order correlation functions show up clearly in the second-order correlation functions, making coincidence counting a valuable tool in quantum optics. The correlation functions have been written above to describe time correlations. *Spatial* correlation functions $g^{(1)}(r)$ and $g^{(2)}(r)$ are defined in a similar manner.

8.3 Applications of entangled beams

In addition to tests of Bell-type inequalities, the other archtypal application of entangled beams is *ghost imaging* or *correlated-photon imaging*. Consider the setup shown in figure 8.1. A pump beam enters a nonlinear crystal, producing signal and idler beams. One beam (call it the signal) is sent through (or reflected off of) an object that is to be imaged, and then is directed to a so-called bucket detector. A bucket detector can be thought of as a single pixel, with no spatial resolution: it can only detect whether or not a photon was received by the detector, but does not record any information as to where it struck the detector surface. As a result, no image can be constructed from the signal detections. The idler is directed to a multi-pixel detector (for example a charge-coupled device (CCD)), which is capable of recording spatial information. However, the idler never encounters the object, so its detection is also insufficient to create an image. In other words, the first-order spatial correlation functions $g^{(1)}(r)$ at either detector are insufficient to reconstruct an image.

However, if coincidence counting is used to measure the second-order correlation function $g^{(2)}(r)$ between the two detectors, then the image emerges in a plot of $g^{(2)}(r)$

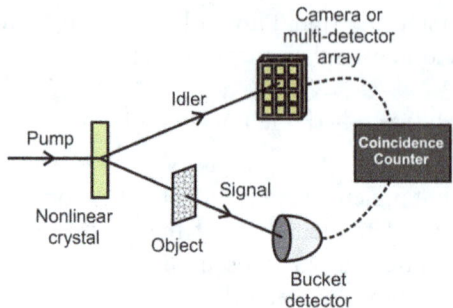

Figure 8.1. Ghost imaging setup. A laser beam provides pump photons to induce spontaneous parametric down conversion in a nonlinear crystal. The resulting signal photon encounters the object to be viewed before reaching a bucket detector. The bucket detector has no spatial resolution; it only registers whether or not a photon arrived. The idler photon, without ever encountering the object, reaches a multi-pixel CCD camera or array of detectors, which is capable of resolving spatial location. A coincidence counter registers a count only when both detectors fire within a very short time window of each other. The resulting coincidence rate measures the second-order correlation function. Although neither detector alone is capable of forming an image of the object, the image reappears in the correlation function.

versus position r! Although data from neither detector contains the image, the correlations between the two data sets does. The image is in a sense spread in a non-local manner between the two branches of the apparatus; this spooky non-local behavior is why the name 'ghost' imaging became common.

Initially, it was believed that entanglement was required for ghost imaging, but it is now known that classical correlations between two beams are sufficient to create a ghost imaging. The difference is that entangled beams can in principle produce perfect image contrast, while classically-correlated beams always produce reduced contrast. The contrast can be given a quantitative meaning. When this is done, it is found that entangled-beam ghost imaging can produce 100% contrast, while the maximum contrast in a classically-correlated beam is $\frac{1}{\sqrt{2}} \approx 71\%$. This is a common feature of many applications of two-beam correlations: classical correlation often mimics the results of entangled versions, but with contrast reduced by a factor of at least $\sqrt{2}$.

In the classical version of ghost imaging, a light beam from a classical thermal source (a hot filament, for example) is split by a beam splitter. Each photon then has equal amplitude to follow each branch of the apparatus. This ensures that not only the *mean* intensity is the same in both beams, but that the *fluctuations* about the mean are identical as well. It is the correlation between the fluctuations that lead to ghost image formation.

Correlated-photon imaging has been shown to have some resistance to the effects of turbulence and aberration under some circumstances [14–26], and applications for it have been proposed, for example, in medical endoscopy where a small and cheap single-pixel camera can be used inside the body, with high-quality images being formed through correlations with a high-resolution camera that is retained outside the body. The use of ghost imaging with entangled photons of opposite orbital angular momentum has been shown to greatly enhance edge contrast in imaging of phase objects (objects which give different phase shifts to light passing through different points). By sending *both* beams through a sample there may be some advantages for specialized forms of microscopy, including resolution beyond the Abbé limit of classical optics.

As mentioned in chapter 4, entangled light beams with anticorrelated orbital angular momentum values can be used in a ghost-imaging light setup to identify the object and reconstruct its rotational symmetries. As was also mentioned in that chapter, OAM-entangled beams are also useful for high precision measurements of rotation rates and rotation angles. Beams entangled in OAM or in polarization can also be used to implement various types of quantum cryptography schemes.

Quantum cryptography, or quantum key distribution (QKD) is a means for two agents, usually called Alice and Bob, to arrive at a shared encryption key, while making sure that no eavesdropper (usually called Eve) is able to gain significant information about the key without being discovered. Alice and Bob may be separated by a large distance, and there may be no way to stop Eve from tampering with their communications, but the goal is to use the rules of quantum mechanics in order to make sure they know when she is acting. One common way to do this is by means of entangled optical beams. Here we sketch the basic idea, assuming the simplest case of polarization-entangled photon pairs. The method relies on the ability to measure photon polarization, which can be done many ways. For example, passage of light through birefringent materials such as quartz will displace the polarization components away from each other, causing them to exit the crystal at different points. Or a polarizing beam splitter can be used to reflect one polarization, while transmitting the other.

Alice wants to send a binary-encoded message to Bob, and she wants to encrypt it using a second binary string as the encryption key. The encrypted message is then the base-2 sum of the original message and the key. As long as the key is completely random, the encrypted message will be completely random as well, ensuring that there are no patterns in the digits that can be used to break the code. If Bob knows the key then he can decode the message by simply adding the key to it a second time, restoring the original message. Alice and Bob need to agree on this random key, while keeping it secure from eavesdroppers. Suppose a third party (which could even be the eavesdropper herself!) uses parametric down conversion to produce entangled signal-idler pairs. The polarization of each photon is random, if measured, but the polarizations of the two are always opposite: if one is horizontal in a given basis, the other is found to be vertical. Alice and Bob make polarization measurements, but they each switch back and forth randomly between two measurement bases: on a given measurement either one could be using the horizontal–vertical (HV) basis or the diagonal-antidiagonal (DA) basis (figure 8.2). When they choose different bases, their results are completely uncorrelated, but when both use the same basis there should be perfect anti-correlation. After their measurements, they tell each other which basis they used for each measurement, but not the outcome of the measurements. They then discard those measurements on which they used different bases. This should leave perfectly anticorrelated results, so that they can use them to agree on a cryptographic key for encoding secret measurements. For example, they could

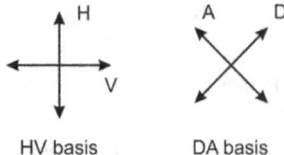

Figure 8.2. Two polarization bases, at 45° to each other. Because $\sin 45° = \cos 45°$, each vector in one basis has equal projections onto both vectors of the other basis. Therefore, if a photon is prepared with a polarization in one basis, then a measurement will give completely random results in the other basis.

agree beforehand that when Alice measures H or D the key bit will be 0, and when she has V or A the bit will be 1. Because of correlation, Bob will be able to reconstruct the same key Alice is using.

However, Eve could try to obtain the key by intercepting some of the photons on their way to Alice and Bob. She could measure their polarizations and then replace them with identical photons of her own making. However she doesn't know which basis is used on any given trial, so she must guess. Each time she guesses wrong, measuring in a different basis from the one used by Alice and Bob, she randomizes their results and destroys the correlations that they expect to find. This is because each state of one basis has equal projections onto both states of the other basis. The resulting drop in correlation is easily detected. While most of the photon pairs are used for the key, a small portion of them are retained for a security check. Alice and Bob send each other their measurement results for this subset, in order for them to compare the results and test the level of correlation. If the correlation level is below that expected after accounting for errors caused by random noise in the detectors and photon losses en route, then they conclude that an eavesdropper was acting. In this case, they abandon the communication channel and seek another one. On average, Eve will introduce an error on $\frac{1}{4}$ of the trials she tampers with. The procedure described above is called the *Eckert or E91 protocol* [27], and is an entangled version of the earlier *BB84 protocol* [28].

The procedure above can be generalized to other variables besides polarization. For example, orbital angular momentum (OAM) may be used [29]. Whereas polarization is a binary variable, the number of possible OAM values is unlimited (at least in principle, if not in practice). This means that large numbers of bits can be encoded into a single OAM-carrying photon, as opposed to the single bit obtained from polarization. For example, if the apparatus is capable of producing photons with any OAM value between $l = 1$ and $l = n$, then $\log_2 n$ bits of the key can be generated from a single photon, where the logarithm is base two. This not only speeds up key-generation rates, it also can be shown to enhance security [30]. When SPDC produces entangled states, the output can be expanded in terms of entangled Bessel or Laguerre–Gauss states. The problems, though, are that the QKD procedure becomes more complicated to carry out with OAM than with polarization, and transmission over large distance is difficult because OAM is fragile in the presence of atmospheric turbulence.

Besides discrete variables like OAM and polarization, QKD has been carried out using continuous variables as well, such as the field quadratures discussed in section 3.3. The idea is that each participant chooses randomly to measure one quadrature of a coherent state, and the results are kept for the trials on which both make the same choice. When one participant measures a given quadrature, the uncertainty in that phase space direction is diminished, decreasing the uncertainty in the other participant's measurement of the same quadrature, effectively squeezing the state, see section 3.3. The results should then be highly correlated. On the other hand, measurements of opposite quadratures will exhibit much weaker correlations, since when one direction is squeezed the other direction must be expanded. Aside from the different choice of measurement variable, the general idea of the key-generating protocol is then similar to that of the discrete case.

For the discrete case, single pairs of entangled photons are used to generate each key bit; if one of the photons is absorbed or scattered, then no key is generated for that trial. Even in the best of cases, the key generation is slow, and the distance over which it works is limited because of photon losses. For the continuous case the situation is different. Strong beams with many photons may be used, since the entire squeezed or coherent state acts as a single quantum state. Loss of a small portion of the beam does not introduce serious errors (at least up to a point), and in order to transmit over a larger distance all that is needed is to use a beam with higher intensity. The trade-off is that the states are harder to generate and entangle, and security is harder to maintain.

For more detailed discussion of quantum cryptography, [6] is a good place to start. More extensive discussions of entangled beam applications in areas such as quantum imaging and quantum communication may be found in [13, 31]. As far as the entangled beams themselves go, entangled Laguerre–Gauss beams are very well-studied, because most work using entangled OAM in areas like quantum cryptography have been carried out with them. But entangled beams of other kinds have also received attention, including entangled Bessel [32], Ince–Gauss [33], and Airy beams [34].

Bibliography

[1] Bell J S 1966 On the Problem of hidden variables in quantum mechanics *Rev. Mod. Phys.* **38** 447
[2] Bell J S 1964 On the Einstein–Podolsky–Rosen paradox *Physics* **1** 195
[3] Clauser J, Horne M, Shimony A and Holt R 1969 Proposed experiment to test local hidden-variable theories *Phys. Rev. Lett.* **23** 880
[4] Clauser J F and Horne M A 1974 Experimental consequences of objective local theories *Phys. Rev.* D **10** 526
[5] Clauser J F and Shimony A 1978 Bell's theorem. Experimental tests and implications *Rep. Prog. Phys.* **41** 1881
[6] Nielsen M A and Chuang I L 2011 *Quantum Computation and Quantum Information: 10th Anniversary Edition* (Oxford: Oxford University Press)
[7] Jaeger G 2007 *Quantum Information: An Overview* (Berlin: Springer)
[8] Jaeger G 2009 *Entanglement, Information, and the Interpretation of Quantum Mechanics* (Berlin: Springer)
[9] Bell J S 1987 *Speakable and Unspeakable in Quantum Mechanics* (Cambridge: Cambridge University Press)

[10] Mermin N D 1990 *Boojums All the Way through: Communicating Science in a Prosaic Age* (Cambridge: Cambridge University Press)
[11] Shih Y 2011 *An Introduction to Quantum Optics: Photon and Biphoton Physics* (Boca Raton, FL: CRC Press)
[12] Ou Z-Y J 2007 *Multi-Photon Quantum Interference* (Berlin: Springer)
[13] Simon D S, Jaeger G and Sergienko A V 2017 *Quantum Metrology, Imaging, and Communication* (Berlin: Springer)
[14] Steinberg A M, Kwiat P G and Chiao R Y 1992 Dispersion cancellation in a measurement of the single-photon propagation velocity in glass *Phys. Rev. Lett.* **68** 2421
[15] Steinberg A M, Kwiat P G and Chiao R Y 1993 Measurement of the single-photon tunneling time *Phys. Rev. Lett.* **71** 708
[16] Franson J D 1989 Bell inequality for position and time *Phys. Rev. Lett.* **62** 2205
[17] Bonato C, Sergienko A V, Saleh B E A, Bonora S and Villoresi P 2008 Even-order aberration cancellation in quantum interferometry *Phys. Rev. Lett.* **101** 233603
[18] Bonato C, Simon D S, Villoresi P and Sergienko A V 2009 Multiparameter entangled-state engineering using adaptive optics *Phys. Rev.* A **79** 062304
[19] Simon D S and Sergienko A V 2011 Correlated-photon imaging with cancellation of object-induced aberration *J. Opt. Soc. Am.* B **28** 247
[20] Simon D S and Sergienko A V 2009 Spatial-dispersion cancellation in quantum interferometry *Phys. Rev.* A **80** 053813
[21] Cheng J 2009 Ghost imaging through turbulent atmosphere *Opt. Exp.* **17** 7916
[22] Zhang P, Gong W, Shen X and Han S 2010 Correlated imaging through atmospheric turbulence *Phys Rev.* A **82** 033817
[23] Li C, Wang T, Pu J, Zhu W and Rao R 2010 Ghost imaging with partially coherent light radiation through turbulent atmosphere *Appl. Phys.* B **99** 599
[24] Meyers R E, Deacon K S and Shih Y 2007 A new two-photon ghost imaging experiment with distortion study *J. Mod. Opt.* **54** 2381
[25] Dixon P B *et al* 2011 Quantum ghost imaging through turbulence *Phys. Rev.* A **83** 051803(R)
[26] Chan K W C *et al* 2011 A theoretical analysis of quantum imaging through turbulence *Phys. Rev.* A **84** 043807
[27] Ekert A K 1991 Quantum cryptography based on Bell's theorem *Phys. Rev. Lett.* **67** 661
[28] Bennett C H and Brassard G 1984 Quantum cryptography: Public key distribution and coin tossing *Proc. IEEE Int. Conf. on Computers, Systems, and Signal Processing, Bangalore* **175**
[29] Groblacher S, Jennewein T, Vaziri A, Weihs G and Zeilinger A 2006 Experimental quantum cryptography with qutrits *New J. Phys.* **8** 75
[30] Cerf N J, Bourennane M, Karlsson A and Gisin N 2002 Security of quantum key distribution using d-Level systems *Phys. Rev. Lett.* **88** 127902
[31] Kolobov M I (ed) 2007 *Quantum Imaging* (Berlin: Springer)
[32] McLaren M, Agnew M, Leach J, Roux F S, Padgett M J, Boyd R W and Forbes A 2012 Entangled Bessel-Gaussian beams *Opt. Exp.* **20** 23589
[33] Krenn M, Fickler R, Huber M, Lapkiewicz R, Plick W, Ramelow S and Zeilinger A 2013 Entangled singularity patterns of photons in Ince-Gauss modes *Phys. Rev.* A **87** 012326
[34] Wei D, Liu J, Yu Y, Wang J, Gao H and Li F 2015 Generation of twin Airy beams with a parametric amplifier *J. Phys. B: At. Mol. Opt. Phys.* **48** 245401

IOP Concise Physics

A Guided Tour of Light Beams
From lasers to optical knots
David S Simon

Chapter 9

Optical knots and links

9.1 From knotted vortex atoms to knotted light

In this chapter, we briefly discuss a fairly new area of research that expands the intersection of beam optics and topology, opening up many new possibilities for future study. The goal here is to give the flavor of current work; the interested reader can then explore more deeply the original references at the end of the chapter.

Consider figure 9.1. Part (a) of the figure shows the interference pattern between two plane waves of the same frequency, tilted at a small angle from each other. The remaining parts show progressively more plane waves added to the superposition. It can be seen that as the number of waves added to the superposition increases, the pattern steadily gains more complex structure. More generally, Fourier analysis guarantees that any continuous spatial pattern on a finite region can be generated by a superposition of a sufficiently large set of plane waves of different direction, phase, amplitude, and frequency.

In the first plot, there are lines of zero intensity (the dark blue regions), where complete destructive interference occurs. These are called *nodal lines*. We have seen nodal lines occur before in the earlier chapters. For example, Laguerre–Gauss beams with non-zero topological charge always have a nodal line along the axis of the beam. In the Laguerre–Gauss case, this nodal line is also a *vortex line* (the phase circulates in a vortex-like pattern around the line), a *singularity* (the phase is singular on the axis), and a *topological defect* (the lines of constant phase collide discontinuously at the axis). In this chapter, we will treat all of these terms as interchangeable, since the cases of interest will fall into all of these categories at once.

As the number of plane waves increases beyond two, the nodal regions in the plane reduce down from lines to points and these points form progressively more complex structures. When the two-dimensional planes shown here are propagated along the z-axis to sweep out three dimensions, these nodal points trace out curves. Some of those nodal curves may extend off to infinity, but others may bend back and twist around in complicated ways, possibly returning to the initial plane many times.

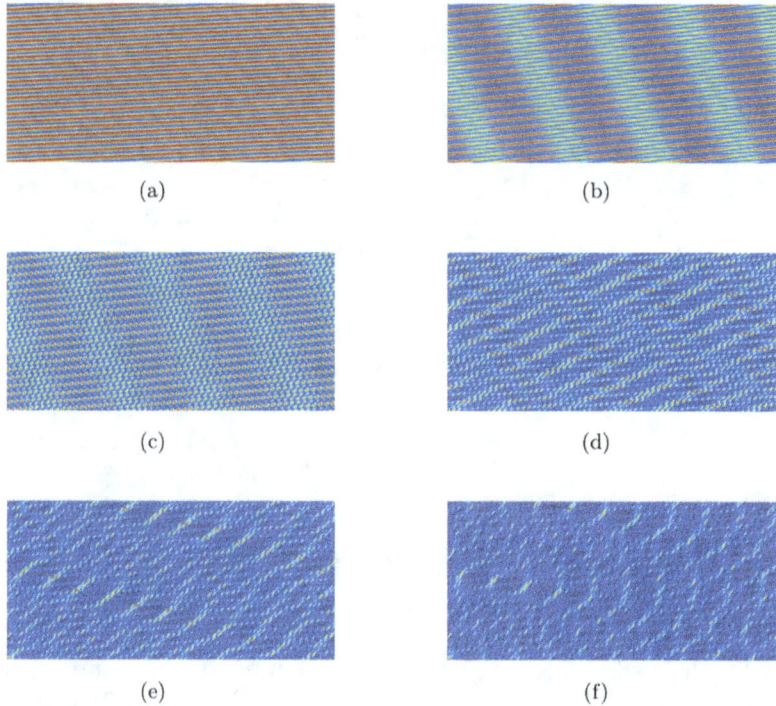

Figure 9.1. Interference of multiple plane waves: (a) two waves, (b) three waves (c) four, (d) six, (e) eight, and (f) ten. Each additional wave is rotated 0.2 radians from the previous one.

There is nothing to prevent the nodal lines from joining back on themselves to form closed loops, or from becoming knotted. Nor does anything prevent them from linking with other nodal lines or forming complicated braid patterns. In fact all of these things can actually be made to occur. More picturesquely, this looping and knotting of nodal curves can be described as filaments of darkness threading themselves through a light field. These knotted and linked threads of darkness can become highly complex and can be used to physically embody topological structures long studied by mathematicians [1–6]. In figure 9.2, 100 plane waves of the same amplitude and wavelength, but each propagating in a random direction, are added; extremely complicated filament structures can be seen to appear, and when the plane is propagated through three dimensions these structures lead to a complex tangle of braided and linked curves.

Knot theory, and related subjects such as braid theory, form a highly developed area of mathematics that has had a long and fruitful interaction with physics. Academic study of knots initially began with George Tait and William Thomson (Lord Kelvin) in the late 19th century, who developed an early model of the atom. In their conception, atoms were formed by knotted vortex lines in the luminiferous ether. Atoms of different elements were distinguished by being knotted in different fashions, and the stability of different chemical elements was protected by their different topological classes. This very appealing idea, however, could not survive

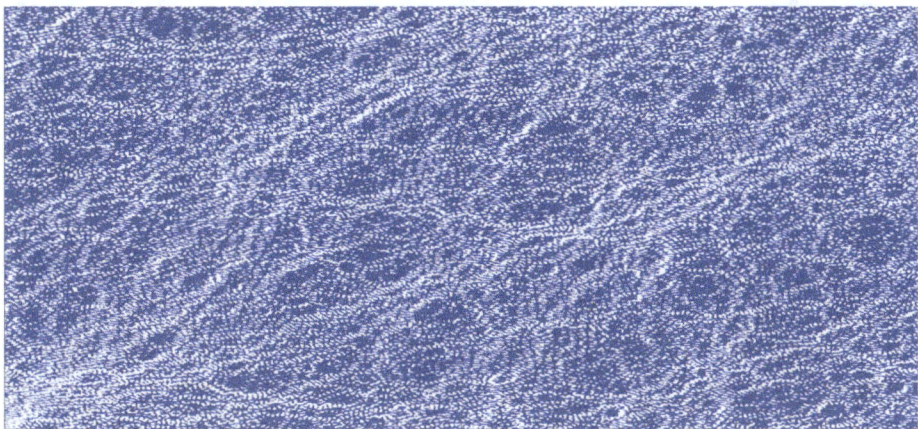

Figure 9.2. The interference pattern produced by adding 100 plane waves in randomly chosen directions. The distribution used to generate the angles was the uniform distribution on the interval $(0, 2\pi)$.

the developments of quantum mechanics and relativity. But knot theory became a well-developed branch of topology, beginning with Tait's initial work on classification of knots. In recent decades knot theory has resurfaced in physics, playing a role in a wide range of areas including Bose–Einstein condensates, superstring theory, and the study of topological defects in cosmology. An entertaining account of the work of Tait and Kelvin on ether knots can be found in chapter 8 of [7]. For an introductory survey of knot theory with emphasis on physical applications, see [8]. In the next section, we focus on a few recent developments involving knots and braids in optics.

9.2 Knotted vortex lines

The question of whether arbitrary nodes and links could be formed on demand using optical nodal lines has recently received a great deal of attention. In the following, we only consider static nodal curves, whose shapes do not change over time as the beams forming them propagate. Thus, each knot stays within some fixed region of finite size within the overlap of the interfering beams, and does not propagate with any of the beams.

It should also be noted that here we discuss knotted vortex lines. But it is also possible to make the electric and magnetic field lines themselves form knots and to thereby implement topological structures such as the well-known Hopf fibration [9] with them. Furthermore, light shaping techniques exist which allow the individual beams to be bent and knotted into arbitrarily-shaped curves [10–14]. These methods have a number of applications for guiding and trapping micron-size particles.

Returning to nodal curves, examples of some of the knotted vortex structures that can be produced are shown in figure 9.3. While those of figure 9.3(a) are simulated, the knots in figure 9.3(b) were extracted from experimental data.

The production of these knotted vortex structures is made possible by the use of spatial light modulators. The spatial light modulator (SLM) in recent years has been

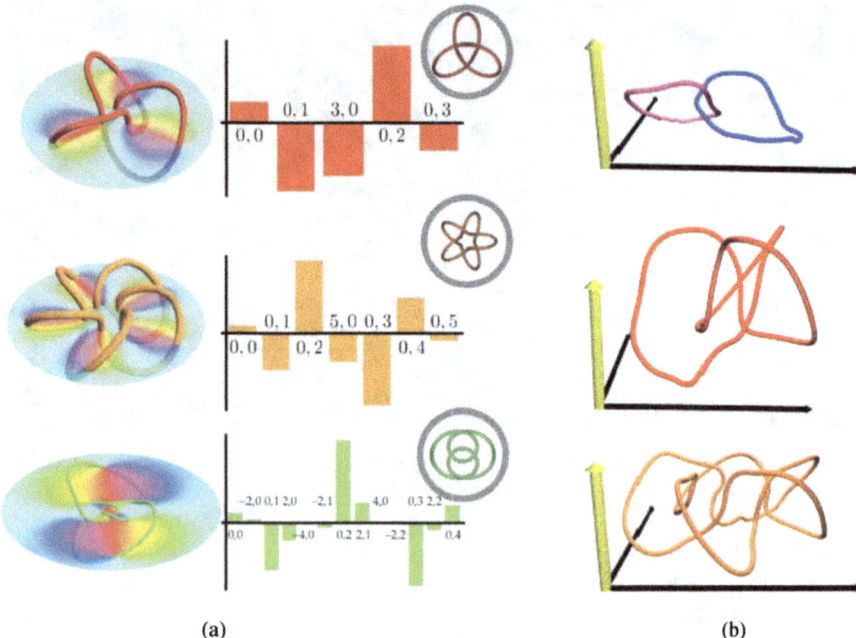

Figure 9.3. (a) Calculated examples of optical vortex knots produced by superpositions of Laguerre–Gauss beams. The histogram gives the amplitude of each Laguerre–Gauss mode, with the modes labelled by their (l,p) values. The knots displayed are known as (from top to bottom) trefoil, cinquefoil, and figure-eight knots. On the left, the tubular shape gives the shape of the vortex line, while the colors on the disk represent the phase near the vortex. (Figures reproduced from [6].) (b) Experimental realizations of knotted vortex lines, reconstructed from phase measurements. The dark central region on the axis of a Laguerre–Gauss or higher-order Bessel beam is bent around and reconnected to itself to form each link. The three figures represent (from top to bottom) two linked circles (a Hopf link), a trefoil knot, and a cinquefoil knot. (Figures reproduced from [5].)

almost as essential a tool as the laser in optics research, as well as being in common use for overhead projector systems and other devices. It consists of a set of cells or pixels, each formed by a liquid crystal, similar to those in many computer monitor screens. The optical properties of the crystal in each pixel can be altered by an electric current, which in turn is controlled by a computer. The phase and amplitude changes added to light reflecting off each pixel can therefore be altered as desired, in order to simulate complex optical patterns. For example, the SLM can be used to simulate a forked diffraction grating of the kind used to produce Laguerre–Gauss beams. In the current context, a computer-controlled SLM is used to create the images of multiple light beams of desired properties, moving at the correct angles to produce the sought-after interference pattern. The appropriate superposition of different beams needed to produce the pattern is calculated numerically, and then the grating pattern and phase shifts needed to implement it are imprinted on the SLM. An example can be seen in the histograms of figure 9.3(a), which shows the amplitudes of the various Laguerre–Gauss beams that are superposed to produce the displayed vortex knots.

Figure 9.4. Two portions of a knot can cross each other in two distinct ways. A binary digit, 0 or 1, can be associated with each type of crossing.

Quantum-mechanically entangled pairs of optical vortex lines have been produced experimentally, and it has been speculated that similar entangled vortex lines of magnetic or other fields may play a role in phenomena such as superfluidity and Bose–Einstein condensation [15]. A possible future application of entangled optical knots is in quantum computing: rather than storing information in entangled binary states, the information would be stored in entangled knots. Different types of knot structures would represent different data values, so that many binary digits could be stored in a single knot. For example, suppose the knot is given an orientation (a preferred direction to travel along the vortex). Then when the vortex line crosses over itself it can cross in two ways, as shown in figure 9.4. The orientation arrow on the upper strand can rotate either clockwise or counterclockwise in order to align with the orientation of the lower strand. These two possibilities can be taken to represent 0s and 1s. So the list of over-crossings and under-crossings can thereby providing a string of binary digits that can be arbitrarily long for a sufficiently complicated knot. More generally, knots have associated with them a number of topological invariants (linking number, Alexander polynomial, HOMFLY polynomial, etc; see [8]) whose values can be used to represent the stored data. Initial work has already been done [16] on the experimental measurement of the tangling in optical vortex knots, which could provide a means of reading the data stored in such a manner. The idea of using topological states, particularly braided states, for quantum computing has been widely investigated in recent years, a field known as *topological quantum computing* [17–19].

Although the study of optical knots and braids is a relatively new area of research, it is clear from this brief sample that a number of intriguing results have already been obtained and that the potential for exciting new developments and applications in the new future is vast.

Bibliography

[1] Berry M V and Dennis M R 2001 Knotted and linked phase singularities in monochromatic waves *Proc. R. Soc.* A **457** 2251

[2] Berry M V and Dennis M R 2001 Knotting and unknotting of phase singularities: Helmholtz waves, paraxial waves and waves in 2 1 spacetime *J. Phys. A: Math. Gen.* **34** 8877

[3] Leach J, Dennis M R, Courtial J and Padgett M J 2004 Knotted threads of darkness *Nature* **432** 165

[4] Dennis M R 2003 Braided nodal lines in wave superpositions *New J. Phys.* **5** 134

[5] Dennis M R, King R P, Jack B, O'Holleran K and Padgett M J 2010 Isolated optical vortex knots *Nat. Phys.* **6** 118
[6] Padgett M J, O'Holleran K, King R P and Dennis M R 2011 Knotted and tangled threads of darkness in light beams *Contemp. Phys.* **52** 265
[7] Wertheim M 2011 *Physics on the Fringe* (New York: Walker Publishing)
[8] Kauffman L H 2013 *Knots and Physics* 4th edn (Singapore: World Scientific)
[9] Irvine W T M and Bouwmeester D 2008 Linked and knotted beams of light *Nat. Phys.* **4** 716
[10] Seldowitz M A, Allebach J P and Sweeney D W 1987 Synthesis of digital holograms by direct binary search *Appl. Opt.* **26** 2788
[11] Soifer V A (ed) 2002 *Methods for Computer Design of Diffractive Optical Elements* (Hoboken, NJ: Wiley)
[12] Whyte G and Courtial J 2005 Experimental demonstration of holographic three-dimensional light shaping using a Gerchberg-Saxton algorithm *New J. Phys.* **7** 117
[13] Gerke T D and Piestun R 2010 Aperiodic volume optics *Nat. Photonics* **4** 188
[14] Rodrigo J A, Alieva T, Abramochkin E and Castro I 2013 Shaping of light beams along curves in three dimensions *Opt. Exp.* **21** 20544
[15] Romero J, Leach J, Jack B, Dennis M R, Franke-Arnold S, Barnett S M and Padgett M J 2011 Entangled optical vortex links *Phys. Rev. Lett.* **106** 100407
[16] Romero M J, Leach J, Jack B, Dennis M R, Franke-Arnold S, Barnett S and Padgett M J 2010 *Frontiers in Optics 2010/Laser Science XXVI* OSA Technical Digest (CD) paper FTuG5 (Washington DC: Optical Society of America)
[17] Freedman M, Kitaev A, Larsen M J and Wang Z 2003 Topological quantum computation *Bull. Am. Math. Soc.* **40** 31
[18] Nayak C, Simon S H, Stern A, Freedman M and Das Sarma S 2008 Non-Abelian anyons and topological quantum computation *Rev. Mod. Phys.* **80** 1083
[19] Simon S 2010 Quantum computing ... with a twist *Phys. World* **35**

A Guided Tour of Light Beams
From lasers to optical knots
David S Simon

Chapter 10

Conclusion

The topics discussed in the earlier chapters are just a small portion of the much larger subject of beam-like waves. There are many additional directions and generalizations that could be discussed. Here, just a few will be briefly mentioned.

All of the beams discussed in this book have acoustic analogs. For example, self-accelerating acoustic beams have been studied and used to direct sound along curved trajectories [1, 2]. Trapping of particles with acoustic Bessel beams has been demonstrated [3].

These beams also, of course, have analogs beyond the optical part of the electromagnetic spectrum. For example, radio waves with non-zero OAM have been studied [4, 5]. Self-accelerating electromagnetic waves have also been studied in curved space-times [6].

Through most of this book, we have ignored the polarization of light beams. In the paraxial approximation this is usually a safe thing to do, but that is no longer true in non-paraxial situations, such as for strongly focused beams where the polarizations may no longer be perpendicular to the propagation axis. Similarly, evanescent waves (the near-field, non-propagating part of the wave) may also be non-transverse.

It can be interesting to look at polarization for other reasons as well: it is possible to design beams in which the polarization varies from point to point within the transverse plane in a non-trivial manner. For example, the polarization may be radially outward from the axis at each point (figure 10.1(a)) or may rotate azimuthally (figure 10.1(b)). Such beams with non-trivial spatially-dependent polarization structures are called *vector beams*.

Also, many of the beams we have discussed have singularities on the axis. Singular solutions of the Maxwell equations also exist that are *not* beam-like. The study of singular solutions in general (beam-like or not) is called *singular optics*, and there is an enormous literature in this area as well [7, 8].

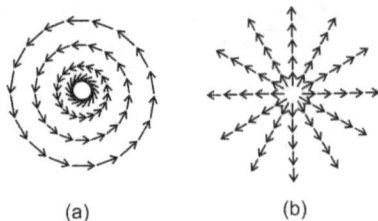

Figure 10.1. Examples of vector beams, in which the polarization direction (represented by the arrows) varies from point to point within the beam. In (a), the polarization is azimuthal, in (b) it is radial. In both cases, there is a singular point on the axis of the beam.

New developments related to optical beams appear on a regular basis. For example, as the writing of this book was being completed, a fundamentally new type of optical vortex was reported [9], in which a toroidal vortex similar to a smoke ring surrounds a central bright pulse. The phase on the ring rotates not around the beam axis, but around the circle at the center of the toroid, an effect not seen before in optics.

From this sampling, it is clear that the study of directed energy beams is extremely rich and diverse, and can serve as an exciting playground for exploration in a number of different fields. It is also still young, having really taken off in just the past 20 years or so, which means it will still be fertile ground for new discoveries and new applications for decades to come.

Bibliography

[1] Bar-Ziv U, Postan A and Segev M 2015 Observation of shape-preserving accelerating underwater acoustic beams *Phys. Rev. B* **92** 100301(R)
[2] Zhao S, Hu Y, Lu J, Qiu X, Cheng J and Burnett I 2014 Delivering sound energy along an arbitrary convex trajectory *Sci. Rep.* **4** 6628
[3] Baresch D, Thomas J L and Marchiano R 2016 Observation of a single-beam gradient force acoustical trap for elastic particles: acoustical tweezers *Phys. Rev. Lett.* **116** 024301
[4] Tamburini F, Mari E, Sponselli A, Thidé B, Bianchini A and Romanato F 2012 Encoding many channels on the same frequency through radio vorticity: First experimental test *New J. Phys.* **14** 033001
[5] Edfors O and Johansson A J 2012 Is orbital angular momentum (OAM) based radio communication an unexploited area? *IEEE Trans. on Antennas and Propagation* **60** 1126
[6] Bekenstein R, Nemirovsky J, Kaminer I and Segev M 2014 Shape-preserving accelerating electromagnetic wave packets in curved space *Phys. Rev. X* **4** 011038
[7] Soskin M S and Vasnetsov M V 2001 Singular optics *Prog. Opt.* **42** 219
[8] Dennis M R, OHolleran K and Padgett M J 2009 Singular optics: optical vortices and polarization singularities *Prog. Opt.* **53** 293
[9] Jhajj N, Larkin I, Rosenthal E W, Zahedpour S, Wahlstrand J K and Milchberg H M Spatiotemporal optical vortices *Phys. Rev. X* **6** 031037

IOP Concise Physics

A Guided Tour of Light Beams

David S Simon

Appendix

Mathematical reference

In this appendix, a number of useful results related to the integrals and special functions used in the text are collected for easy reference. More extensive listings of such results can be found in standard mathematical reference books, such as [1].

A.1 Gaussians

The transverse profile of most optical beams are either a pure Gaussian or a Gaussian modulated by another function. As a result, Gaussian integrals occur frequently in this area. More generally, they are ubiquitous throughout math and physics, in areas ranging from quantum field theory to statistics. Here we record some basic results related to Gaussian integrals.

In one-dimension the basic Gaussian integral is:

$$\int_{-\infty}^{\infty} e^{-ax^2} dx = \sqrt{\frac{\pi}{a}}. \tag{A.1}$$

Some related integrals are:

$$\int_{-\infty}^{\infty} x^{2n} e^{-ax^2} dx = \frac{\Gamma\left(n + \frac{1}{2}\right)}{a^{\frac{2n+1}{2}}} \tag{A.2}$$

$$\int_{-\infty}^{\infty} x^{2n+1} e^{-ax^2} dx = 0 \tag{A.3}$$

$$\int_{-\infty}^{\infty} e^{-ax^2 + bx} dx = \sqrt{\frac{\pi}{a}} e^{\frac{b^2}{4a}}. \tag{A.4}$$

By replacing $a \to -ia$ in the last result, we find the further integral:

$$\int_{-\infty}^{\infty} e^{iax^2 + ibx} dx = \sqrt{\frac{i\pi}{a}} e^{\frac{-ib^2}{4a}}. \tag{A.5}$$

In n-dimensions, we have the generalized integral

$$\int e^{-\frac{1}{2}\sum_{i,j} A_{ij} x_i x_j} e^{\sum_{i=1}^{n} B_i x_i} d^n x \sqrt{\frac{(2\pi)^n}{\det(A)}} \exp\left(\frac{1}{2} \boldsymbol{B}^T A^{-1} \boldsymbol{B}\right), \tag{A.6}$$

for any vector $\boldsymbol{B}$ and any symmetric, positive-definite matrix A.

A.2 Laguerre polynomials

The associated or generalized Laguerre polynomials $L_p^\alpha(x)$ are polynomials of order p that satisfy the Laguerre differential equation: $x\, y'' + (\alpha + 1 - x) y' + py$, where the primes denote derivatives with respect to x. p is an integer and α a real number. The first two polynomials are

$$L_0^\alpha(x) = 1, \qquad L_0^\alpha(x) = 1 - x + \alpha, \tag{A.7}$$

from which all the other polynomials may be obtained via the recurrence relation

$$L_{p+1}^\alpha(x) = \frac{1}{p+1}\Big((2p + 1 + \alpha - x) L_p^\alpha(x) - (p + \alpha) L_{p-1}^\alpha(x)\Big). \tag{A.8}$$

Additional recurrence relations obeyed include

$$p\, L_p^\alpha(x) = (p + \alpha) L_{p-1}^\alpha(x) - x L_{p-1}^{\alpha+1}(x) \tag{A.9}$$

$$L_p^\alpha(x) = L_p^{\alpha+1}(x) - L_{p-1}^{\alpha+1}(x), \tag{A.10}$$

as well as the addition formula:

$$L_p^{\alpha+\beta+1}(x + y) = \sum_{n=0}^{p} L_n^\alpha(x) L_{p-n}^\beta(y). \tag{A.11}$$

They can be obtained from the Rodrigues relations

$$L_p^\alpha(x) = \frac{x^{-\alpha} e^x}{p!} \frac{d^p}{dx^p}(e^x x^{p+\alpha}) = \frac{x^{-\alpha}}{p!}\left(\frac{d}{dx} - 1\right)^p x^{p+\alpha}, \tag{A.12}$$

or the generating function

$$\sum_{p=0}^{\infty} L_p^\alpha(x)\, t^p = \frac{1}{(1-t)^{\alpha+1}} e^{-tx/(1-t)}. \tag{A.13}$$

The ordinary Laguerre polynomials, $L_p(x) = L_p^0(x)$ obey

$$L_p(x) = \sum_{k=0}^{p} \binom{p}{k} \frac{(-1)^k x^k}{k!}, \tag{A.14}$$

$$L_{-p}(x) = e^x L_{p-1}(-x), \tag{A.15}$$

$$\frac{d}{dx}L_p(x) = \left(\frac{d}{dx} - 1\right)L_{p-1}(x), \tag{A.16}$$

and, for $\alpha = k =$ integer:

$$L_p^k(x) = (-1)^k \frac{d^k}{dx^k} L_{p+k}(x). \tag{A.17}$$

The associated Laguerre functions obey orthonormality relation:

$$\int L_p^\alpha(x) L_q^\alpha(x) e^{-x} x^\alpha \, dx = \frac{(p+\alpha+1)!}{p!} \delta_{pq}. \tag{A.18}$$

From this it is straightforward to show that the Laguerre–Gauss (LG) functions of chapter 4 are orthonormal:

$$\int u_p^l(x)\left[u_q^m(x)\right]^* d^2x = \delta_{pq}\delta_{lm}. \tag{A.19}$$

These functions form a complete set, so any optical state may be decomposed into a sum of LG modes. It may be convenient sometimes to separate the LG functions into radial and angular parts:

$$u_{lp}(r, z, \phi) = K_p^l(z) R_p^l(r, z) e^{-il\phi}, \tag{A.20}$$

where

$$\begin{aligned} K_p^l(z) &= \frac{C_p^{|l|}}{w(z)} e^{i(2p+|l|+1)\arctan(z/z_R)} \\ R_p^l(r, z) &= \left(\frac{\sqrt{2}r}{w(z)}\right)^{|l|} e^{-r^2/w^2(r)} L_p^{|l|}\left(\frac{2r^2}{w^2(r)}\right) e^{-ikr^2 z/(2(z^2+z_R^2))}. \end{aligned} \tag{A.21}$$

A.3 Bessel functions

There are several types of Bessel functions. The type used in chapter 5 is called the *Bessel function of the first kind*, which is written as $J_\nu(x)$ and is a solution to Bessel's differential equation,

$$x^2 J_\nu''(x) + x J_\nu'(x) + (x^2 - \nu^2) J_\nu(x) = 0. \tag{A.22}$$

The primes denote derivatives. ν can be any real number, although we will mostly need the case where it is an integer. Bessel functions often appear in systems that have cylindrical symmetry or that are described in cylindrical coordinates; as such, they arise naturally in the study of optical beams, which have cylindrical symmetry about the propagation axis. There are many references for Bessel functions, including any book on math methods in physics (such as [2]); but if you really want to go deep into the gory details, try [3]. Bessel functions were first used by Daniel Bernoulli (1700–1782), and then rediscovered by Friedrich Wilhelm Bessel (1784–1846) while studying the many-body problem in gravitational physics.

Some of the basic properties of the Bessel function are listed for easy reference in this appendix. In the following formulas, the index will be written as ν when it can be any real number and will be denoted as n in relations that are true only when the index is an integer.

The Bessel functions obey several recurrence relations, such as

$$J_{\nu-1}(x) + J_{\nu+1}(x) = \frac{2\nu}{x} J_\nu(x) \tag{A.23}$$

$$J_{\nu-1}(x) - J_{\nu+1}(x) = 2J'_\nu(x). \tag{A.24}$$

They obey an orthonormality condition (for $\nu > -1/2$),

$$\int_0^\infty J_\nu(ar) J_\nu(br) r \, dr = \frac{1}{a} \delta(a - b), \tag{A.25}$$

and a Rodriguez relation,

$$\exp\left[\frac{x}{2}(t - t^{-1})\right] = \sum_{n=-\infty}^\infty J_\nu(x) t^n. \tag{A.26}$$

The exponential function on the left is called a generating function, because the Bessel functions can be generated from it by taking derivatives and then setting $t = 0$ afterward:

$$J_n(x) = \frac{1}{n!} \frac{d^n}{dt^n} \left(e^{\frac{x}{2}(t - t^{-1})} \right) \bigg|_{t=0} \tag{A.27}$$

There are many different representations for the Bessel functions, including a series expansion,

$$J_\nu(x) = \sum_{s=0}^\infty \frac{(-1)^s}{s! \Gamma(\nu + s + 1)} \left(\frac{x}{2}\right)^{\nu+2s}, \tag{A.28}$$

and (for integer n) an integral representation,

$$J_n(x) = \frac{1}{\pi} \int_0^\pi \cos(n\phi - x \sin \phi) d\phi. \tag{A.29}$$

A more general integral representation that also works for non-integer ν is

$$J_\nu(x) = \frac{1}{\pi} \int_0^\pi \cos(\nu\phi - x \sin \phi) d\phi - \frac{\sin \nu\pi}{\pi} \int_0^\infty e^{-\nu\phi - x \sinh \phi} d\phi. \tag{A.30}$$

Some useful Bessel-related integral relations include:

$$\int_0^{2\pi} e^{-ix\cos\theta} d\theta = 2\pi J_0(x) \tag{A.31}$$

$$\int_0^\infty J_\nu(bx) dx = \frac{1}{b} \tag{A.32}$$

$$\int_0^u x^n J_{n-1}(x) dx = u^n J_n(u). \tag{A.33}$$

Different positive integer orders are linked by derivatives,

$$J_n(x) = (-x)^n \left(\frac{1}{x} \frac{d}{dx} \right)^n J_0(x), \tag{A.34}$$

($n = 1, 2, 3, \ldots$), while positive and negative integer orders are related to each other by

$$J_{-n}(x) = (-1)^n J_n(x). \tag{A.35}$$

One other fact that is useful is that the Bessel functions form a complete set: any other function can be expanded as a sum or integral over Bessel functions. This is analogous to Fourier analysis, where a function is expanded in sine and cosine functions, or in terms of complex exponentials. An example of this is expressed by the Sommerfeld integral:

$$\frac{e^{ikr}}{r} = i \int_0^\infty dk_r \frac{k_r}{k_z} J_0(k_r r) e^{ik_z z}, \tag{A.36}$$

where $k_r^2 + k_z^2 = k^2 = (\frac{\omega}{c})^2$. The left side of this equation is a radially-expanding spherical wave, while the right side is its expansion in terms of cylindrical zero-order Bessel waves of different radial momenta.

A.4 Hermite polynomials

Hermite polynomials were first studied by Laplace (1810) and Chebyshev (1859) before being rediscovered by Charles Hermite in 1864. They often occur as solutions to differential equations in Cartesian coordinates, and in particular appear in the solutions to the quantum harmonic oscillator equation. They are relevant to optics because they form a basis for the higher-order modes of lasers (see section 7.2).

The first few Hermite polynomials are given by

$$H_0(x) = 1 \tag{A.37}$$

$$H_1(x) = 2x \tag{A.38}$$

$$H_2(x) = 4x^2 - 2 \tag{A.39}$$

$$H_3(x) = 8x^3 - 12x \tag{A.40}$$

$$H_4(x) = 16x^4 - 48x^2 + 12. \tag{A.41}$$

If the first two polynomials are known, the others can be generated via the recurrence relation

$$H_{n+1}(x) - 2xH_n(x) + 2nH_{n-1}(x) = 0. \tag{A.42}$$

Many explicit representations of these polynomials exist, including:

$$H_n(x) = (-1)^n e^{x^2} \frac{d^n}{dx^n} e^{-x^2} \tag{A.43}$$

$$= \frac{(-2i)^n}{\sqrt{\pi}} \int_{-\infty}^{\infty} e^{-t^2} t^n e^{2ixt} dt \tag{A.44}$$

$$= \sum_{k=0}^{[n/2]} \frac{(-1)^k n!}{k!(n-2k)!} (2x)^{n-2k}. \tag{A.45}$$

They obey the orthonormality relation

$$\int_{-\infty}^{\infty} e^{-x^2} H_m(x) H_n(x) dx = 2^n n! \sqrt{\pi} \delta_{mn}, \tag{A.46}$$

and can be obtained from a generating function:

$$H_n(x) = \left(\frac{d}{dt}\right)^n \bigg|_{t=0} G(x, t), \tag{A.47}$$

where

$$G(x, t) = \sum_{n=0}^{\infty} \frac{H_n(x)}{n!} t^n = e^{2xt - t^2}. \tag{A.48}$$

Their polynomials and their derivatives at $x = 0$ obey

$$H_{2n}(0) = \frac{(-1)^n (2n)!}{2^n n!} \tag{A.49}$$

$$H_{2n+1}(0) = 0 \tag{A.50}$$

$$\frac{d^n H_n(x)}{dx^n} \bigg|_{x=0} = 2^n n!. \tag{A.51}$$

The polynomials are solutions of the differential equations:

$$H'_n(x) = 2nH_{n-1}(x) \quad \text{and} \quad H''_n(x) - 2xH'_n(x) + 2nH_n(x) = 0 \tag{A.52}$$

The integral

$$\frac{1}{\sqrt{2\pi}}\int_{-\infty}^{\infty} e^{ixy}e^{-y^2/2}H_n(y)dy = i^n e^{-x^2/2}H_n(x) \qquad (A.53)$$

is occasionally useful, as is the 'addition formula'

$$\sum_{k=0}^{n}\binom{n}{k}H_{n-k}(x\sqrt{2})H_k(y\sqrt{2}) = 2^{n/2}H_n(x+y). \qquad (A.54)$$

Since the Hermite polynomials form a complete set, linear combinations of them may be used to construct any other square-integrable function on the real line. For example, trigonometric functions and hyperbolic functions may be written in terms of them as:

$$\sinh 2x = e \sum_{k=0}^{\infty} \frac{1}{(2k+1)!}H_{2k+1}(x) \qquad (A.55)$$

$$\cosh 2x = e \sum_{k=0}^{\infty} \frac{1}{(2k)!}H_{2k}(x) \qquad (A.56)$$

$$\sin 2x = e^{-1}\sum_{k=0}^{\infty}(-1)^k\frac{1}{(2k+1)!}H_{2k+1}(x) \qquad (A.57)$$

$$\cos 2x = e^{-1}\sum_{k=0}^{\infty}(-1)^k\frac{1}{(2k)!}H_{2k}(x). \qquad (A.58)$$

Bibliography

[1] Gradshteyn I S, Ryzhik I M, Jeffrey A and Zwillinger D 2000 *Table of Integrals, Series, and Products* 6th edn (London: Academic)

[2] Arfken G, Weber H and Harris F E 2012 *Mathematical Methods for Physicists: A Comprehensive Guide* 7th edn (London: Academic)

[3] Watson G N 1995 *A Treatise on the Theory of Bessel Functions* 2nd edn (Cambridge: Cambridge University Press)